Seismic Isolation for Earthquake-Resistant Structures

Advances in Earthquake Engineering

Objectives

The objectives of this series are to provide clear accounts of both basic and applied research in the various fields of earthquake engineering with particular reference to earthquake resistant analysis and the design of structural systems.

The series consists of books concerned with state-of-the-art developments in earthquake engineering and as such comprises several volumes every year covering the latest developments and applications. Each volume is composed of authored works or edited volumes of several chapters written by leading researchers in the field.

The scope of the series covers almost the entire spectrum of earthquake engineering and as such the following topics will be discussed: engineering seismology, strong ground motions and site effects, seismic hazard evaluation and design earthquake loads, soil-structure interaction, numerical methods in earthquake engineering, stochastic analysis methods, principles of earthquake resistant design, reinforced concrete structures, steel structures, masonry and masonry infill structures, historical buildings and monuments, bridges, earth and concrete dams, underground and lifeline structures, storage tanks, silos and other industrial structures, offshore structures, seismic isolation and control, vulnerability and risk assessment of structural systems, repair and retrofit, seismic code regulations and case studies in earthquake engineering.

Series Editors

Seismic Isolation for Earthquake-Resistant Structures

Petros Komodromos
Intelligent Engineering Systems Laboratory
Department of Civil and Environmental Engineering
Massachusetts Institute of Technology

WIT*PRESS* Southampton, Boston

Petros Komodromos
Massachusetts Institute of Technology

Published by

WIT Press
Ashurst Lodge, Ashurst, Southampton, SO40 7AA, UK
Tel: 44 (0) 238 029 3223; Fax: 44 (0) 238 029 2853
E-Mail: witpress@witpress.com
http://www.witpress.com

For USA, Canada and Mexico

WIT Press
25 Bridge Street, Billerica, MA 01821, USA
Tel: 978 667 5841; Fax: 978 667 7582
E-Mail: infousa@witpress.com
http://www.witpress.com

British Library Cataloguing-in-Publication Data
A Catalogue record for this book is available
from the British Library

ISBN: 978-1-85312-803-5
ISSN: 1361-617X

Library of Congress Catalog Card Number: 00-104027

Printed in Great Britain by Lightning Source, UK.

Contents

Preface xv

1. Introduction 1

1.1 Motivation ... 1
1.2 Outline ... 6

2. Conventional Earthquake Resistance Design and its Limitations 11

2.1 Objectives and Design Philosophy ... 11
2.2 Disadvantages ... 13
2.2.1 Resonance problem for low- to medium-rise structures ... 13
2.2.2 Consequences of conventional earthquake resistance design philosophy ... 17
2.3 Need for a Different Design Approach ... 17

3. Seismic Isolation: An Alternative Earthquake Resistance Design Approach 19

3.1 The Concept ... 19
3.2 Basic Components ... 19
3.3 Advantages and Applications of Seismic Isolation ... 20
3.4 Comparison of a Fixed Supported and a Seismically Isolated Structure ... 22
3.5 Constraints in the Implementation ... 26
3.6 Limitations ... 27
3.6.1 Superstructure characteristics ... 27
3.6.2 Seismic hazard and site characteristics ... 27
3.6.3 Surrounding structures ... 28
3.6.4 Minor lateral loads ... 29
3.7 Factors Which Enable the Use of Seismic Isolation ... 29
3.8 Economics of Seismic Isolation ... 30
3.9 Reasons for the Lack of Extensive Use of Seismic Isolation ... 32

4. Seismic Isolation Devices **35**

4.1 Types of Seismic Isolation Systems 35
4.2 Sliding Isolation Systems 35
4.3 Elastomeric Isolation Systems 37
4.3.1 Lead rubber bearings (LRBs) 39
4.3.2 High damping rubber bearings (HDRBs) 43
4.3.3 Hybrid type: lead high damping rubber bearing (LHDRBs) 44
4.4 Elastomeric Bearings: Mechanical and Physical Characteristics 45
4.4.1 Characteristics 45
4.4.2 Design considerations 49
4.4.3 Available evaluation tests of seismic isolation and energy dissipating devices 58

5. Applications of Seismic Isolation **61**

5.1 Introduction 61
5.2 Seismic Isolation in the United States 62
5.2.1 The Foothill Communities Law and Justice Center 63
5.2.2 The Salt Lake City and County Building 64
5.2.3 The University Hospital of the University of Southern California 65
5.2.4 Seismic retrofit of Oakland City Hall using seismic isolation 66
5.2.5 San Francisco City Hall 68
5.2.6 Los Angeles City Hall 69
5.3 Seismic Isolation in Japan 71
5.3.1 The Computer center of the Tohoku Electric Power Company 72
5.3.2 The C-l Building 72
5.4 Seismic Isolation in New Zealand 73
5.4.1 The William Clayton Building 74
5.4.2 Wellington Central Police Station 75
5.4.3 New Zealand Parliament Buildings 76
5.5 Seismic Isolation in Italy 77
5.6 Observed Performance of Seismically Isolated Structures 78
5.6.1 The University Hospital of the University of Southern California 79
5.6.2 The West Japan Computer Center 81
5.6.3 The Eel River Bridge 85
5.6.4 The Rangitaiki River Bridge at Te Teko, New Zealand 85

6. Practical Considerations for Seismic Isolation **87**

6.1 Seismic Isolation of Buildings 87
6.1.1 Configuration of seismic isolation system 87
6.1.2 Practical issues and problems 88
6.2 Seismic Isolation of Bridges 92

6.2.1 Configuration of seismic isolation system 93
6.3 Maintenance and Management of the Isolation System 96
6.4 Code Requirements 97
6.4.1 Regulations in the United States 97
6.4.1.1 Uniform Building Code 98
6.4.1.2 Codes for seismic isolation of hospitals 108
6.4.1.3 Codes for seismic isolation of existing buildings 109
6.4.1.4 Codes for seismic isolation of bridges 109
6.4.2 Regulations in Japan 109

7. Structural Modeling and Analysis of Seismically Isolated Structures **111**
7.1 Modeling of Response 111
7.2 Natural Modes of Vibration 112
7.3 Linear Systems 117
7.4 Bilinear Systems 120
7.5 Comparative Study of a Linearly and a Bilinearly Isolated Structure 120
7.6 Linearization of Bilinear Behavior 124
7.6.1 Effective stiffness 124
7.6.2 Effective damping 125
7.6.3 Study of the effectiveness of bilinear behavior linearization 127
7.7 Effect of the Degree of Non-Linearity 131
7.8 Effect of Superstructure Stiffness 134
7.9 Modeling of Seismically Isolated Bridge Structures 136
7.10 Ductility of Seismically Isolated Structures 138

8. A Preliminary Design Methodology **139**
8.1 Introduction 139
8.2 A Feasibility Study for the Use of Seismic Isolation 140
8.3 Determination of Peak Ground Acceleration 141
8.4 Design of the Superstructure 142
8.5 Selection of a Suitable Isolation System 143
8.6 Preliminary Design of the Isolation System 144
8.6.1 Estimation of the maximum allowable displacement 145
8.6.2 Estimation of yield force F_y of isolators 145
8.6.3 Determination of required effective stiffness 146
8.6.4 Determination of plan size dimensions of isolators 152
8.6.5 Determination of required total rubber thickness 155
8.6.6 Design corrections and adjustments 155
8.7 Issues of Concern at the Final Analysis Stage 158

9. The Future of Seismic Isolation **163**
9.1 Summary 163
9.2 Future of Seismic Isolation 168

Appendix A: Linear Elastic Systems 173
A.1 Single Degree of Freedom Systems ... 173
A.2 Multi Degree of Freedom Systems ... 175
A.2.1 Modal analysis ... 176
A.2.2 Newmark method ... 180

Appendix B: Bilinear Systems 181
B.1 Single Degree of Freedom Systems ... 181
B.2 Multi Degree of Freedom Systems ... 185

References 189

Index 199

Acknowledgements

Firstly, I would like to express my gratitude to Prof. Dimitri Beskos, who suggested the publication of this book. I am extremely grateful to him for the endless help and support he has given me since I first met him during my undergraduate studies at the University of Patras in Greece. I would also like to express my gratitude and thanks to my undergraduate professors in Greece, in particular my advisor Prof. Michael Fardis, and one of my most favorite professors, Prof. Stavros Anagnostopoulos, for their encouragement and support throughout my studies. Together with Prof. Beskos, they have been my mentors and have significantly contributed to my education and to the fulfillment of my academic goals. Their advice has been invaluable.

Next, I would like to thank my advisor during my Master of Science studies at MIT, Prof. Jerome Connor, for his enthusiasm and encouragement during my graduate studies at MIT. I am also grateful to all my professors during my graduate studies, both here at MIT and at the University of California at Berkeley, for the valuable knowledge they have shared with me.

I would also like to thank Prof. Carlos Brebbia and the WIT Press staff, especially Mr. Lance Sucharov and Miss Clare Garcia, for their help and cooperation.

Finally, I would like to thank my family and friends in Cyprus and Greece for believing in and supporting me. It is indeed a blessing to have them all emotionally so close to me. Their unconditional love, emotional support and encouragement have always been my driving force in pursuing my goals and fulfilling my dreams. In particular, it is difficult to find words to express my love and gratitude to my parents, Ioanni and Androula, whose love, encouragement, and understanding have always been a constant source of motivation and strength for me. Finally, I would also like to specifically thank my sister Marina, who has been next to me during all the hard times I went through, my best friend George Houliotis, and the sweetest girl I have ever met, Marina Cocconi.

Preface

Ground motion, due to earthquake excitation, often induces disastrous disturbances that severely affect structures and their contents. Conventional earthquake-resistant design focuses on the strengthening of structures to resist these disturbances and avoiding structural collapse, whilst little attention is given to the prevention of damage. Using this design approach, it is almost impossible to construct completely 'earthquake-proof' structures which are of reasonable cost and aesthetically acceptable.

Seismic isolation, an alternative innovative performance-based design approach, may be used to minimize earthquake induced loads, and to mitigate or reduce resulting damage in low- to medium-rise buildings. This book explores the use of this alternative design approach to address the destructive consequences of strong earthquake excitations. The basic idea of seismic isolation is rather simple: reduction of the earthquake-induced inertia loads by shifting the fundamental frequency of the structure out of the dangerous-for-resonance range, and concentration of the deformation and energy dissipation demands at the isolation system, which is designed for this purpose.

The need for the method is discussed, given the limitations of conventional earthquake-resistant design, and is followed by a detailed description of its characteristics, advantages and limitations. The capability of a seismic isolation system to decouple a structure from the damaging effects of ground acceleration, and thus reduce seismic forces on the structure, is demonstrated. Current isolation systems and their components are described in detail. The advantage of using a hybrid system is emphasized. The performance of seismically isolated structures is evaluated, and examples of such structures and their response during an earthquake are presented. Finally, a preliminary design methodology based on seismic isolation is proposed. It involves several design stages, including the design of the isolation system and that of the superstructure, and the estimation of critical parameters. The latter will be used in the final analysis to determine the most appropriate design. The potential of the method and its establishment as a powerful earthquake-resistance design approach are also assessed.

Recent seismic events have shown the necessity of imposing performance requirements in order to avoid damage and ensure safety. Conventional design methods aim at avoiding structural collapse and resulting casualties at the expense of damage. The lack of awareness that a structure designed using conventional earthquake resistant methods is not 'earthquake proof' prevents the use of advanced earthquake-resistant methods such as seismic isolation. The aim of this book is to help the practicing engineer or student to understand the concept of seismic isolation and how it can be utilized in an effort to successfully encounter strong earthquakes and to mitigate their consequences.

Petros I. Komodromos
Cambridge, Massachusetts

Dedicated to my parents

Chapter 1

Introduction

1.1 Motivation

Earthquakes are natural threats which, to a degree depending on their magnitude, can be disastrous. For structures in high seismicity regions, earthquake loading is considered the most significant and possibly the most destructive external load, particularly for low- to medium-rise buildings, which constitute the overwhelming majority of buildings worldwide. In contrast, wind loads usually dominate the design of high-rise buildings, even in highly seismic areas. Today, technological advances allow us to control the consequences and hazards of severe earthquakes. Unfortunately, advances in design and construction technology are based partially on experience gained after the occurrence of severe earthquakes with extensive damage and loss of human life. However, such advances are only rarely utilized in earthquake resistance design, especially with regard to common building projects.

In the dominant conventional earthquake resistance design approach, safety measures are emphasized while measures to reduce the resultant damage are given less attention. Although methods of analysis, design and construction of earthquake-resistant structures have substantially improved in the last decade, severe earthquakes still result in undesirable damage, even in the case of structures designed according to the most conservative and rigorous seismic codes. For example, the objective of the latest Uniform Building Code concerning earthquake resistant design, is explicitly stated in Section 1626 as follows: *"The purpose of the earthquake provisions herein is primarily to safeguard against major structural failures and loss of life, not to limit damage or maintain function"* (Paragraph 1626.1, UBC-97, [111]). In essence, the seismic codes are life-safety oriented, and do not aim at preventing property damage and functionality disruption. However, high damage levels and loss of functionality, especially of critical operations, are no longer acceptable considering the higher performance-demand implied by the technological advances of our times, and the high-cost equipment housed in structures that must be protected. Thus, alternative design and construction methods must be employed to address earthquake dangers to structures and their contents.

Motivated by the necessity for an innovative design approach to address the problem of resisting strong earthquake excitations without damage, this book explores seismic isolation, which is also known as base isolation. In contrast to most traditional design methods, which aim at increasing the earthquake resistance capacity of structures, seismic isolation aims at reducing the seismic loads induced by earthquake excitations. This design approach is very elegant since it is in step with the aim of an engineer intelligently to seek avoidance of problems, rather than to deal with their consequences.

Earthquake resistance design can be viewed as a "supply and demand" problem. Conventional earthquake resistance design aims at increasing the "supply" by providing higher strength to equate at some level the "demand" in order to avoid structural collapse and human casualties. In most cases it is impossible to provide the necessary strength to equate the demand required in order to avoid damage. In addition, in the effort to increase the "supply" side, i.e. the strength, the "demand" side may also increase due to increased seismic forces. In contrast, seismic isolation aims at reducing the "demand" side of the "supply and demand" equation, by reducing the seismic forces.

The idea of isolating a structure from earthquake motions of its supporting ground were proposed as early as the end of the nineteenth century. The first known proposal for seismically isolating a structure was suggested more than a century ago, in Japan. Other proposals followed during the next few decades, but none was actually implemented. It is only during the last few decades that seismic isolation has been established as a method of achieving earthquake protection for a structure, and has been practically implemented in built projects. The latest developments in earthquake engineering and seismology, the advances in computer and software engineering, and the innovations and improvements in the design and manufacture of elastomeric bearings and dampers have enabled the utilization of seismic isolation to mitigate earthquake hazards. During the last few years there has been an increasing use of seismic isolation to design and build earthquake-resistant structures, as well as seismically to upgrade existing structures, especially those of historical value.

The main objective of earthquake engineering should be to design and build a structure in such a way that damage to the structure and its contents during an earthquake is minimized. This can be achieved by reducing the inter-story displacements and floor accelerations of the structure under strong seismic motion. The primary causes of damage to structural and nonstructural components of a building and its contents are the deformations and floor accelerations of the structure (Figure 1.1a).

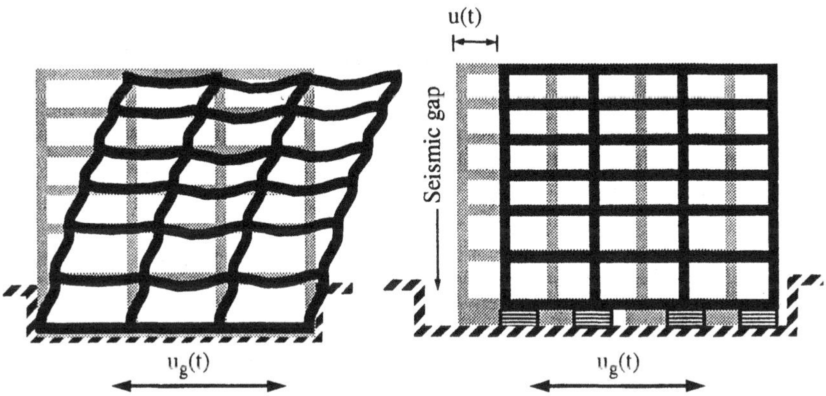

Figure 1.1: Deformation patterns of a fixed supported and a seismically isolated building during an earthquake excitation.

The inter-story displacements or the story shear strains cause damage to the structural elements of the building, as well as to its nonstructural elements and the equipment that interconnects floors. The floor accelerations cause damage to the contents of the structure which in some cases, such as nuclear plants, may be devastating. In order to minimize the inter-story displacements, the structure must be stiffened, which, however, leads to amplification in the building of the earthquake ground accelerations with high floor accelerations. In contrast, by reducing the stiffness of the structure the floor accelerations can be reduced but the inter-story displacements are increased due to the structural flexibility. Seismic isolation can be used to reduce simultaneously the inter-story displacements and floor accelerations of a building.

Conventional earthquake resistance design methods aim at avoiding structural collapse by allowing structural damage, i.e. inelastic deformations, to occur in order to absorb the earthquake energy. In contrast, seismic isolation does not allow the earthquake energy to enter the structure by providing flexibility which decouples it from the earthquake motion. In addition, the energy dissipation is achieved in the isolation system, which is designed to serve such a purpose, and not within the structure. A seismically isolated structure has a fundamental frequency that is much lower than the fundamental frequency of the corresponding fixed-supported structure and the usual predominant frequencies of a typical earthquake. This is achieved by mounting the structure on a set of isolators (Figure 1.1(b)) that provides low horizontal stiffness and, consequently, shifts the fundamental frequency of the structure to much lower values.

Although a seismically isolated structure as an entity may move much more than a fixed supported structure, most deformations occur at the isolation level. The superstructure remains essentially undeformed (Figure 1.1(b)), moving like a rigid body, which prevents damage to its structural and nonstructural components. It is the isolation system that is deformed, and this can be designed to accept the anticipated maximum displacements. The floor accelerations are, also, significantly reduced compared to the floor accelerations of a fixed-supported structure, which are, usually, significantly amplified by the ground accelerations. The maximum floor accelerations of a seismically isolated structure may be even less than the peak ground acceleration.

A major advantage of using seismic isolation is that, by shifting the fundamental frequency of the structure away from the dangerous-for-resonance range, amplification of the ground accelerations is avoided. The reduction of the magnitudes of the floor accelerations, which is very important for protection of the contents of a structure, is not possible using conventional earthquake resistance design methods. Figure 1.2(a), shows a smoothed typical acceleration spectrum and Figure 1.2b shows a typical smoothed displacement response spectrum.

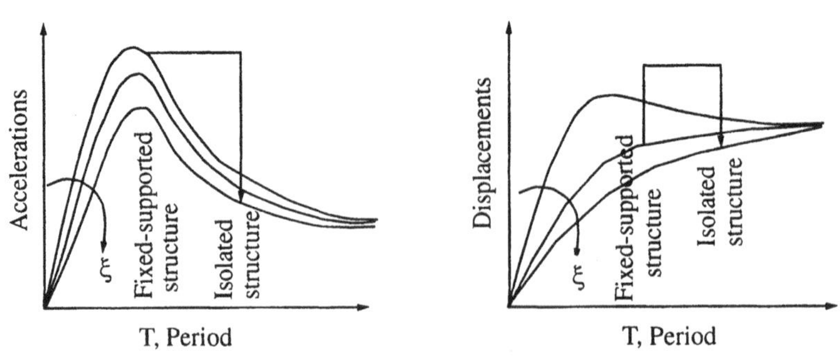

(a) Acceleration response spectrum (b) Displacement response spectrum

Figure 1.2: Typical response spectra.

As is demonstrated by the acceleration spectrum in Figure 1.2(a), resonance can be avoided by increasing the period of a structure, which achieves a substantial decrease in accelerations. Additional damping, which is usually provided by the isolation system, can further reduce the accelerations. Therefore, increasing the fundamental period of a structure, by inserting flexibility at its base, avoids resonance and substantially reduces the accelerations and seismic forces that the structure experiences.

Considering displacements, the displacement response spectrum (Figure 1.2(b)) shows that by increasing the period of the system, the displacements may increase. Increased energy dissipation, which can be provided by the isolation system, or by auxiliary dampers, may control the increased displacements. Although the overall displacements of an isolated structure may be much larger than those of a similar fixed-supported structure, they are concentrated at the isolation level, and the superstructure remains essentially rigid. The inter-story deflections and, consequently, the story shear forces, are significantly lower than those of a similar fixed-supported structure, avoiding damage to the structural and nonstructural components of the structure. Not only the deformations but also the dissipation mechanisms are concentrated in the isolation system, instead of being widespread throughout the structure. The isolation system can be designed to be able to undertake large strains, and can be more easily monitored and repaired than the entire structure.

The major requirement when using seismic isolation is the required seismic gap that must be provided around the structure to allow the free movement and the large displacements that are anticipated at the isolation level during a strong earthquake. The required clearance around the structure is essential for the unimpeded function of seismic isolation and must be ensured for the entire lifetime of the structure. In some cases it may be difficult, or even impossible, to use seismic isolation due to the non-availability of the required clearance around the structure to be seismically isolated.

Seismic isolation can also be used to isolate only a part of a structure, or even certain valuable contents or critical and expensive equipment housed in a building, rather than to isolate the entire structure. Sensitive equipment or the valuable contents of a building can be protected from earthquake excitations by mounting them on devices which provide vibration isolation. An entire floor inside a building can be isolated, also, by mounting it on isolation devices. Such isolators are ideal for hospitals, nuclear reactors and other critical facilities. However, this book concentrates on the seismic isolation of structures and not the specialized seismic isolation of equipment, or contents, housed in a building.

The advantages and limitations of seismic isolation are presented in detail in this book, as well as its potential for use in earthquake-prone areas. Unfortunately, this advanced earthquake resistance design method has not yet been widely established, for reasons which are also identified and discussed. Finally, ways to promote and motivate its wider use are proposed.

Although seismic isolation has been increasingly utilized for structures that house critical facilities, or valuable contents, it has not yet been widely utilized as a seismic protection method for common construction or retrofitting projects. However, this tendency has almost always characterized new inventions, such as cars, radio, televisions and computers. New products are almost always prohibitively expensive for ordinary people at their initial production stages due to the high initial investment and manufacturing costs. In addition, new inventions are always faced with conservatism, especially in the field of civil engineering where risks to human life, in the case of malfunction, cannot be neglected. As the superior performance of seismically isolated structures is observed during violent earthquakes, the method is gaining the necessary reliability for its wider establishment and use. Mass production of isolators and wider use of the method will reduce the cost of seismic isolation to levels where it can be employed in common construction and retrofitting projects.

1.2 Outline

Seismic isolation is an alternative to conventional design methods for dealing with earthquake loadings in relatively high-frequency structures, especially in low- to medium-rise buildings. To place the discussion in its appropriate context and eventually to highlight the potential of seismic isolation, current conventional earthquake resistance design philosophy and its disadvantages and limitations are discussed first, in Chapter 2. Conventional earthquake resistance design approaches aim at avoiding excessive damage during moderate earthquakes only; for strong earthquakes the objective is to avoid structural collapse and consequent casualties, irrespective of whether the structure may need to be demolished once the earthquake is over. This is achieved by providing the structural elements with sufficient strength to avoid their failure, together with mechanisms of inelastic deformations inserted into the structure to dissipate the seismic energy.

Although a properly designed structural system with careful detailing of the structural elements and their connections may save lives, it does not usually protect the structure and its contents. The necessity for an alternative design approach to address the issue of damage is emphasized. Damage is no longer acceptable, especially for buildings housing critical facilities, which should remain operational during and after the earthquake, or for infra-structure such as highway bridges, which need to remain functional after a strong earthquake. In addition, historic buildings of great cultural value and museums that house valuable exhibits cannot be left unprotected. Similarly, buildings that house corporation headquarters, high-tech and -cost industrial plants, and communication and control centers need to be protected, since damage of their contents or interruption of their operation will be devastating for the economy. The severe consequences of powerful earthquakes on such structures justifies the use of more advanced earthquake resistance design approaches to mitigate the damage. However, even residential buildings can be safeguarded by such advanced design approaches if the latter become cost effective and feasible for use by the general public to protect their properties and belongings.

Seismic isolation is discussed in detail in Chapter 3. First, the fundamental features of seismic isolation and its goal, namely to decouple the structure from the damaging effects of ground accelerations, are presented. Chapter 3 includes discussion of the basic components and advantages of seismic isolation as well as a simple example in which a fixed supported structure and a seismically isolated one are compared, to demonstrate the effectiveness of the method. Constraints and limitations in the practical implementation of seismic isolation are also discussed, followed by an account of the factors which enable its development and use. Some economic issues related to seismic isolation are mentioned, for both new and retrofitted structures. Finally, reasons for the lack of extensive use of seismic isolation are discussed.

Chapter 4 describes the most commonly used seismic isolation devices, and in particular sliding systems and elastomeric bearings. The sliding bearings limit the transfer of shear across the isolation interface to a value that depends on the coefficient of friction, the geometry of the sliding surface, and the vertical load. The elastomeric bearings, which are made of thin vulcanized rubber sheets bonded to thin steel plates, are very flexible in the horizontal direction and very stiff in the vertical direction. They introduce high flexibility at the isolation level of the isolated structure, reducing its horizontal stiffness and, consequently, lengthening its fundamental period. The discussion in Chapter 4 focuses on elastomeric bearings, since they are the most commonly used seismic isolation devices.

In particular, elastomeric bearings are designed to be highly elastic in the transverse direction, in order to be able to sustain large deformations without suffering any damage. High vertical stiffness to avoid vertical rocking of the isolated structure is achieved by the use of steel reinforcing steel sheets on which rubber sheets are bonded. The most frequently used elastomeric bearings are lead rubber bearings (LRBs) and high-damping rubber bearings (HDRBs). The former have a solid lead plug inserted, usually in the middle of the bearing, while the latter are manufactured using a certain type of filled rubber compound that provides a high damping capacity. As the elastomeric bearings deform, frictional heat is generated, either by the lead plug or by the high-damping rubber compound, providing an energy dissipation mechanism. The energy dissipation is important in order to dissipate the kinetic energy induced by the ground motion, by converting it to heat, and to damp the building's oscillations. The characteristics of the lead and high-damping rubber bearings are given in Chapter 4, as well as those of a hybrid isolation system. The latter incorporates the advantages of both basic schemes using high-damping rubber bearings combined with lead plugs. The potential of the hybrid system for extensive use in seismic isolation applications is emphasized.

The practical implementation of seismic isolation, related practical problems and maintenance and management issues are discussed in Chapter 5. First, a historical overview and selected practical applications of seismic isolation are presented, followed by information on the observed performance of seismically isolated structures. In particular, applications of seismic isolation in the United States, Japan, New Zealand and Italy are presented, followed by a description of the observed performance of seismically isolated structures under strong earthquake excitations.

In Chapter 6 the practical issue of selecting an appropriate location for the isolation level is discussed, as well as some practical problems that may arise when the isolation system is incorporated into a structure. Both seismically isolated buildings and bridges are considered. First, potential configurations of an isolation system in buildings, and issues that may arise when seismically isolating a building, are presented. Some of the practical problems presented apply not only to buildings but to all types of structures that are seismically isolated.

The discussion then focuses on seismic isolation of bridge structures. The advantages of using seismic isolation for bridge structures, not only to reduce seismic loads but also to control the distribution of seismic and non-seismic lateral forces, are presented. The cost-effectiveness and superior performance of bridges whose superstructures are seismically isolated is emphasized. The elastomeric bearings which are typically used to isolate the bridge's superstructure seismically, act under normal conditions as conventional bearings. In the event of a strong earthquake excitation, they provide the necessary flexibility and energy dissipation mechanism to reduce and dissipate the seismic energy that enters the structure.

Next, general management and maintenance issues with regard to seismically isolated structures are discussed. Finally, pertinent seismic isolation codes and provisions in the United States and Japan are identified and presented. In particular, concerning seismic isolation regulations in the United States, the relevant provisions of the Uniform Building Code (UBC-97, [111]) are explained in detail, and this is followed by a brief description of specific codes for the seismic isolation of hospitals, existing buildings, and bridges.

Chapter 7 describes the structural characteristics of commonly used isolation systems. First, different mathematical models used in the preliminary design stage and the final analysis are discussed. Then, the stiffness and damping characteristics of both linear and bilinear systems are presented and a comparative study of both systems is performed, in order to identify the advantages and disadvantages of each system. The alternative use of a hybrid system, such as one composed of high-damping rubber lead bearings, is proposed, based on the results of the above-mentioned study. The linearization of the bilinear behavior of the isolation system and a study on its accuracy are also included. Finally a study to determine the effect of superstructure stiffness on the behavior of the isolated structure is finally performed.

A preliminary design procedure for lead-rubber bearings is proposed in Chapter 8, using results obtained in the previous chapter. The procedure is based on the maximum acceptable displacement which must be accommodated during the useful life span of the structure. It provides initial values for the dimensions and the configuration of the isolation system which are needed in the final and more detailed analysis. Finally, issues of concern at a more detailed analysis stage are discussed and their importance is emphasized.

Concluding remarks are presented in the last chapter of this book, Chapter 9. In particular, a summary of conclusions is provided and important observations are discussed, and these are followed by an assessment of the potential and future of seismic isolation.

Chapter 2

Conventional Earthquake Resistance Design and its Limitations

2.1 Objectives and Design Philosophy

In high seismicity areas, it is almost impossible to built a conventionally fixed supported low- to medium-rise building of reasonable cost which is also "earthquake proof", against even a severe earthquake. Therefore, emphasis is usually given to the limitation of casualties and the avoidance of structural collapse. Damage, which results from inelastic deformations, is the safety valve to dissipate incident energy due to the earthquake and to reduce, when the structure is properly designed, the probability of structural collapse. Although this design approach has saved many lives in technologically advanced countries where it is properly applied, it has still resulted in extensive damage with severe economic consequences.

The purpose of conventional earthquake resistance design, as defined in most current design codes, is to resist minor earthquakes and prevent all damage, to resist moderate earthquakes with limited non-structural damage, and to prevent structural collapse under a severe earthquake but with acceptance of both structural and non-structural damage. Structural damage refers to damage of structural members, while non-structural damage refers to damage of non-structural parts of the structures, such as architectural facades, partition walls and ceilings. Properly designed conventional earthquake-resistant structures have the strength and ductility to withstand seismic loads and avoid structural collapse, but allow structural damage.

Seismic loads are due to accelerations of the masses of the structure which, for low- to medium-rise buildings, are amplifications of the ground accelerations. Conventional earthquake resistance design is based on reduced seismic loads, smaller than those expected if the structure could withstand them without inelastic deformations; otherwise it would be practically impossible to provide the strength required to avoid inelastic deformations within reasonable economic and architectural costs. This reduction is achieved by providing adequate levels of ductility to structural members so that they can experience inelastic deformations without failure during a major earthquake. The ductility capacity is provided by proper design and construction of the structural members, to yield at particular locations and be capable of withstanding several cycles of reversed load without brittle failure. The energy inserted in the structure due to the earthquake excitation is absorbed through plastic deformations resulting in significant and undesirable damage. This energy absorption mechanism is harmful not only to individual structural members but also to adjacent non-structural elements. The absorption of energy is implicitly taken into consideration in the framework of an elastic analysis by using reduction factors to account for ductility. The seismic capacity of a relatively stiff structure increases when the structure experiences inelastic deformations. This increase is due to the resulting reduction of stiffness which shifts the fundamental frequency of the structure towards the range where seismic effects are less significant.

Most seismic codes are based on design spectra constructed using several earthquake excitations and their corresponding response spectra. A response spectrum is a plot of maximum displacement, velocity or acceleration for many possible single degree of freedom (DOF) systems due to a particular excitation. A design spectrum is usually constructed to envelop the mean value plus a standard deviation of several response spectra. Considering the uncertainties associated with both structural characteristics and the expected earthquake excitation, a design spectrum may be preferable to a response spectrum. Figure 2.1 shows a typical design spectrum.

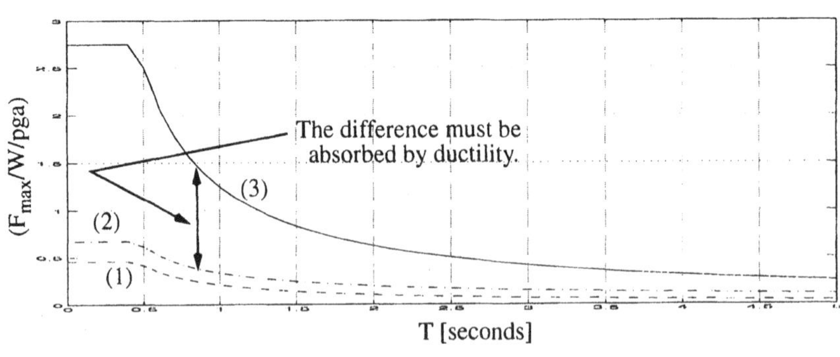

Figure 2.1: Typical design spectrum.

The top curve (3) depicts the force level, according to the Structural Engineers Association of California (SEAOC, [100]), of a fixed supported structure if the structure was designed to withstand the seismic loads elastically. The lowest curve (1) corresponds to the forces which would be used in the design of a fixed structure according to the Uniform Building Code (UBC [108]). Taking into account the safety factors used in the design, the estimated strength of a fixed-supported structure is given by curve (2). The difference between the maximum elastic force and the probable strength must be absorbed by ductility.

2.2 Disadvantages

At the beginning of this discussion it was mentioned that conventional earthquake resistance design has some disadvantages. The most important ones are briefly discussed here.

2.2.1 Resonance problem for low- to medium-rise structures

A serious problem that arises in low- to medium-rise buildings is that their fundamental frequency is in the range of the dominant frequencies of earthquakes. This results in amplification of the ground acceleration causing severe damage to structures and their contents. This disadvantage of conventional earthquake resistance design may be demonstrated through the construction of a response spectrum.

Through a mode superposition analysis, it can be shown that the response of most structures is characterized mainly by motion in the first few natural eigenmodes. The contribution of the fundamental eigenmode is of great importance since it is, at least in most cases, the dominant portion of the structural dynamic response. Most modern seismic codes provide equivalent static procedures based on the contribution of only the fundamental mode.

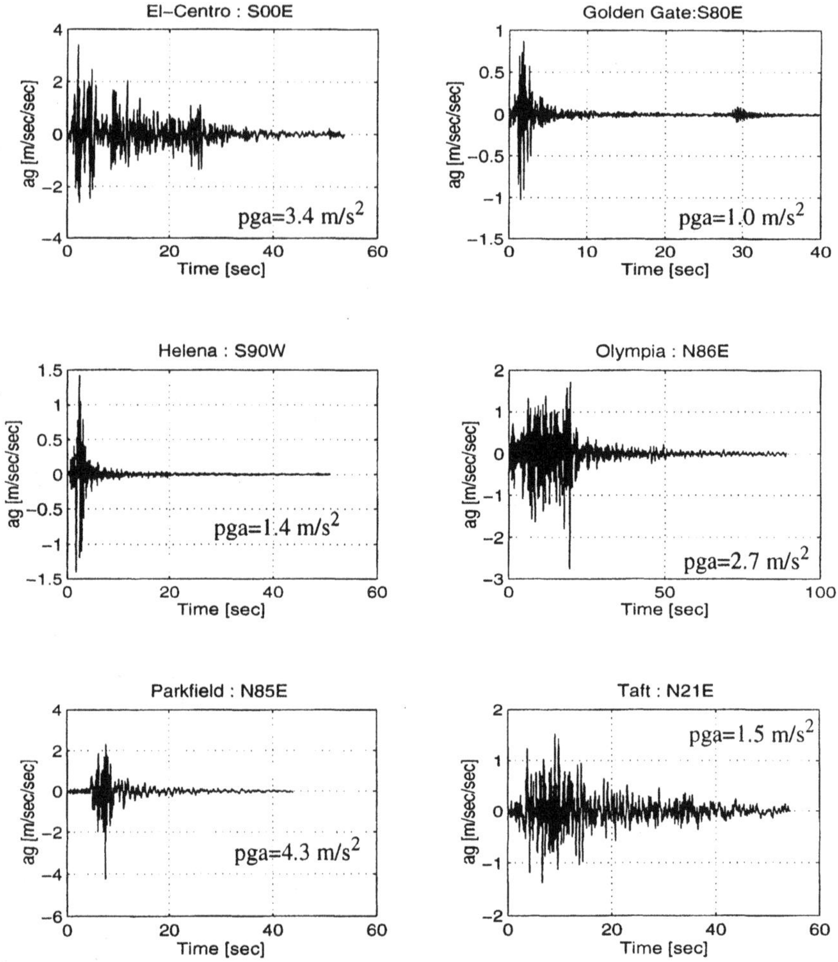

Figure 2.2: Earthquake excitations.

The response spectra of six earthquake excitations, shown in Figure 2.2, have been constructed using the Newmark α–δ method, described in Appendix A. The influence of the fundamental period of a structure on its response is evident from the response spectra. The response spectra of maximum relative displacements and maximum pseudo-accelerations for these excitations, and for damping ratios ranging from 0% to 20%, are plotted in Figures 2.3 and 2.4, respectively.

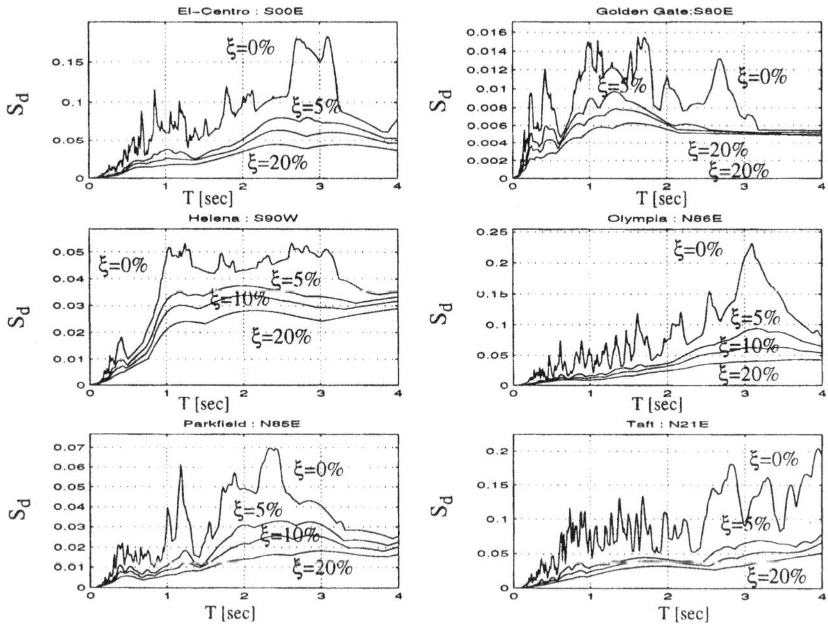

Figure 2.3: Maximum relative displacements S_d.

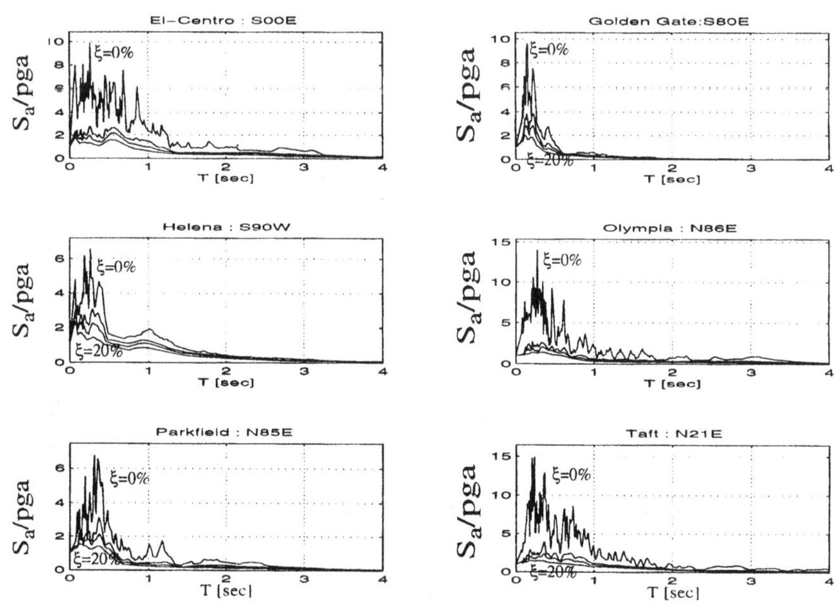

Figure 2.4: Spectral pseudo-accelerations S_a.

Spectral displacement and accelerations are reduced with increasing damping ratio ξ. The relative displacement S_d increases while the spectral acceleration S_a decreases as the natural period increases. Assuming that a typical low-rise building has a fundamental period typically in the range of 0.1 to 0.7 sec, we observe that at its fundamental frequency it is in resonance. To avoid such a situation, the period must be either lengthened, by increasing the flexibility of the structure, or shortened, by increasing the stiffness. An increase in stiffness reduces relative displacements, while an increase in flexibility results in substantial reduction of accelerations.

The effect of the natural period of a structure may be illustrated by considering a mass connected to the ground with an infinitely flexible spring, as shown in Figure 2.5(a). In the case of an earthquake excitation, i.e. support motion, the mass will remain essentially in its initial position, unable to follow the ground motion. This response will lead to approximately zero acceleration and zero force in the spring. However, the relative displacement will equal the ground displacement $u_g(t)$. Consider now a mass connected to the ground by an infinitely rigid spring, as shown in Figure 2.5(b). In this case, the mass will follow the ground motion. This response implies zero relative displacements. However, the acceleration will equal the ground acceleration $a_g(t)$, which may be beyond the acceptable limit.

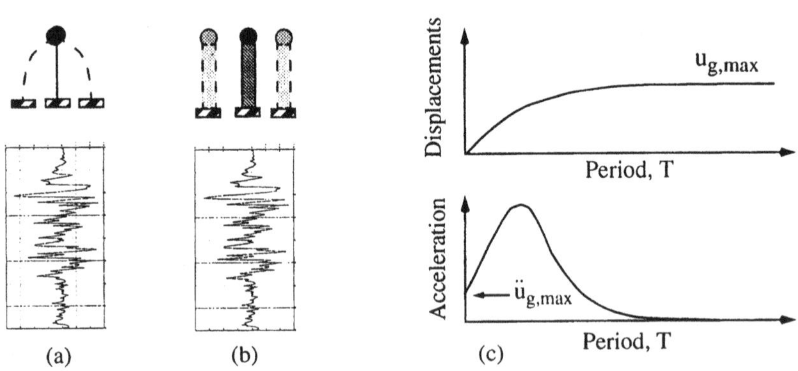

Figure 2.5: Smoothed response spectra of single DOF systems.

It follows from the single mass example that a very flexible structure will have very large inter-story displacements but small accelerations and consequently small shear forces. In contrast, a very stiff structure will have very small inter-story displacements but accelerations equal to the ground accelerations and relatively high seismic forces. Therefore, using the conventional earthquake resistance design approach we cannot reduce both inter-story deflections and floor accelerations in order to avoid damage.

2.2.2 Consequences of conventional earthquake resistance design philosophy

In most cases, it is virtually impossible to construct an earthquake-proof structure. Thus, earthquake resistance design is geared towards human safety and thus the avoidance of structural collapse. The consequences of an earthquake excitation in a conventionally designed structure include structural and non-structural damage, mainly due to inter-story displacements and in particular inter-story drifts (or story shear strains). These are the relative displacements between two subsequent floors, divided by the story height. Acceptable story shear strains are typically less than 1/500, to avoid non-structural damage. Special design and construction details, such as isolation of the facade walls from the surrounding structural members, may allow higher values for story shear strain.

Another consequence is damage of building contents due to the floor accelerations which usually increase with height. In the case of buildings with very sensitive and usually expensive equipment the consequences may be catastrophic. A final but most crucial consequence is the disruption of operation of the damaged structures for repairs. This is disastrous, especially in the case of hospitals and fire stations which must remain fully operational, particularly after a severe earthquake.

2.3 Need for a Different Design Approach

To minimize or eliminate the above consequences and still use the conventional design approach, construction costs must be significantly increased, the architectural design must be aesthetically affected, and several construction methods unsuitable for earthquake resistance must be completely avoided. Even if the structure were to behave elastically under a severe seismic excitation, the resulting stresses would be several times higher than the yield stress of the material, if members of reasonable size were used. This is actually the reason why structural engineers compromise in the conventional earthquake resistance design approach and assume lower seismic loads than those expected. This compromise relies on the ability of the structure, when properly designed and constructed, to provide high levels of ductility, and implicitly accepts the above damage.

In recent years it has been realized that in the case of a severe earthquake the safety of residents is not the only issue of concern. First of all, the conventional earthquake resistance design approach is expensive due to the high cost of repairs. Second, the loss of functionality of structures which have essential facilities which must remain operational, and the protection of very sensitive and expensive equipment, are also very important. Finally, human discomfort during a violent earthquake and the resulting psychological effects which are related to the magnitude of floor accelerations, must also be addressed. Thus, a different design approach is needed, to account for the above-mentioned effects of seismic excitations.

Chapter 3

Seismic Isolation: An Alternative Earthquake Resistance Design Approach

3.1 The Concept

Seismic isolation is an old design idea, proposing the decoupling of a structure or part of it, or even of equipment placed in the structure, from the damaging effects of ground accelerations. One of the goals of seismic isolation is to shift the fundamental frequency of a structure away from the dominant frequencies of seismic excitations and the fundamental frequency of the fixed superstructure. This innovative design approach aims mainly at the isolation of a structure from the supporting ground, usually only in the horizontal direction, in order to reduce the transmission of the earthquake motion to the structure. In most applications it is placed at the base of the structure and is thus referred to as base isolation.

Although three-dimensional isolation systems have also been developed to account for seismic excitation in both the horizontal and vertical directions, they are very complicated and thus difficult to implement practically. In addition, for most structures the vertical earthquake components are not as dangerous as the horizontal ones. The main reason for this is that the majority of structures are designed and built to carry vertical loads and are therefore less vulnerable to additional such loads. Also, the vertical components of ground motion are usually weaker than the horizontal ones.

3.2 Basic Components

Seismic isolation introduces flexibility at the base of the structure in the horizontal direction, shifting its fundamental frequency away from the dominant frequency range of the earthquake. In addition, it also provides an energy dissipation mechanism at the level of isolation, reducing the relatively large relative displacements between the superstructure and the supporting ground. Finally, the seismic isolation system provides either rigidity under minor lateral loads, e.g. wind loads, or an energy dissipation mechanism to suppress the response due to service loads.

3.3 Advantages and Applications of Seismic Isolation

The increased flexibility of the seismically isolated structure is its most important feature and results in the avoidance of resonance and the decrease of accelerations of the structure at different levels. In Chapter 2 it was demonstrated that, while a very flexible structure has very small accelerations and very large relative displacements, a very rigid structure has very small relative displacements but accelerations equal to the ground acceleration. Seismic isolation actually takes advantage of the benefits of both the above-mentioned extreme cases, reducing both inter-story deflections and floor accelerations. Inter-story deflections are reduced because of the rigid body motion of the superstructure. The superstructure is relatively very stiff compared to the inserted flexibility at the isolation level. The reduction of inter-story deflections of the structure minimizes or even eliminates structural or non-structural damage.

In addition, the lengthening of the fundamental period due to the inserted flexibility reduces the accelerations significantly. This is an important achievement, in particular for buildings with very sensitive equipment which must remain operational even after a severe earthquake. Finally, appropriate design of the isolation system may reduce torsional effects due to the eccentricities of the superstructure. This may easily be achieved by coinciding the center of stiffness of the isolation system with the center of mass of the superstructure.

Another advantage of seismic isolation is related to the energy dissipation mechanism. In conventional earthquake design, the energy dissipation mechanism is based on plastic deformations at scattered locations in the structure. In contrast, in the case of seismic isolation the energy dissipation mechanism is concentrated at the isolation level and may therefore be more easily designed, particularly to withstand several inelastic cycles under reversed loading, and may also be monitored to ensure appropriate performance. Note that the design must allow for the easy replacement of the energy dissipation mechanism in the case of malfunction, as will be described in Chapter 6. Since the dissipation of energy of a seismically isolated structure does not depend primarily on the ductility of the superstructure but on the dissipation mechanism of the isolation system, the ductility requirements of the superstructure could ideally be partially released. For the time being, however, considering the limited experience of seismically isolated structures, and the uncertainties associated with the method itself and the nature and frequency of occurrence of the ground excitations as well as other unexpected excitations, such as earthquakes with strong vertical and long period components, the existing ductility requirements should not be released.

Seismic isolation is based on the local insertion of isolators and the provision of a gap to accommodate the relative displacement of the superstructure at the isolation level. It may be applied to both new and existing structures, especially in bridges which are usually founded on common thermal expansion bearings. In such cases, the installation of the isolation system is achieved simply by replacing the common bearings with the isolation ones. Lead rubber bearings are suitable in these cases, as they provide all the features of seismic isolation in one compact unit. In addition, proper sizing of the lead plug cylinder and the bearing may be used in the redistribution of the lateral forces from individual weaker piers of lower seismic capacity to stronger piers and abutments.

Properly dimensioned seismic isolators can be used to control the distribution and the way of transfer of the lateral forces. This feature of seismic isolation may achieve significant cost savings compared as to conventional earthquake strengthening approaches. Recent strong earthquakes and resulting traffic disruption indicate the necessity of seismic upgrading of existing bridges so that they remain operational after such powerful events. Many bridges have been built with insufficient seismic performance requirements. These bridges must be seismically upgraded and seismic isolation is a relatively cheap and effective method of doing this, easily implemented without significant disruption to the operation of the bridge.

Seismic isolation may also be used to retrofit existing structures not appropriately designed for seismic loads, and in particular to rehabilitate historical structures which must maintain their original architectural characteristics. In these cases, instead of increasing the ability of the structure to withstand typical seismic loads for a fixed supported structure, the imposed seismic loads are substantially decreased using seismic isolation. The required ability to withstand seismic forces and the ductility demand capacity of the structure are decreased to appropriate levels.

Seismic isolation is the least disruptive technique since most of the construction work involved is usually confined to the level of isolation, typically at the basement of the building. The loss of functionality of the superstructure is limited, considering that only limited stiffening or strengthening may be required. This is very important for buildings that house essential facilities which must remain operational during seismic retrofitting. In contrast, conventional strengthening of buildings involves extensive modifications at all levels, and the loss of their functionality is unavoidable. For historic structures, where preservation of the architectural characteristics of the superstructure dominates the design, seismic isolation has the least impact aesthetically. This is also due to the fact that seismic isolation reduces the level of seismic forces and there is, therefore, no reason significantly to modify the superstructure to sustain the large seismic forces of a fixed supported one.

In contrast, conventional strengthening techniques modify the structure to make it stronger which in turn attracts higher seismic forces. Unfortunately, practical constraints in some existing buildings, historical ones in particular, do not allow the use of seismic isolation.

Finally, seismic isolation may be used for new structures in combination with methods of construction which are usually not suitable in earthquake resistance design, since the induced seismic forces are substantially decreased without the need for high ductility.

3.4 Comparison of a Fixed Supported and a Seismically Isolated Structure

A three-story structure under both fixed-supported and seismically isolated boundary conditions, as shown in Figures 3.1 and 3.2 respectively, is studied to demonstrate the effectiveness of seismic isolation. The structural characteristics of both models and their responses are summarized in Table 3.1. The latter include maximum inter-story deflections, mass accelerations and story shear forces at each level of both the fixed-supported and the seismically isolated structures.

The analysis is performed assuming the El-Centro excitation, scaled to have a peak ground acceleration (pga) of 0.50 g. It is observed that maximum floor accelerations, inter-story deflections and story shear forces are reduced substantially, even if the energy dissipation mechanism of the isolation system is neglected. For both structures a 5% damping ratio ξ is used, although a much higher value of damping corresponds to a seismically isolated structure. If the relative damping, either hysteretic or viscous, of the isolation system is taken into account, further reductions in the response of the isolated structure result.

Level	Masses	Stiffness	Maximum deflections [mm]		Maximum accelerations [m/sec^2]		Maximum story shears [MN]	
	[Tons]	[MN/m]	Fixed	Isolated	Fixed	Isolated	Fixed	Isolated
Isolation	335	30	---	136.1	---	6.2	---	4.5
1	305	1250	5.7	2.4	6.2	6.3	7.1	3.3
2	285	1125	5.1	1.7	9.4	6.4	5.7	2.1
3	265	1125	2.9	0.8	12.1	6.4	3.3	1.0

Table 3.1: Structural characteristics and maximum response of both systems.

The fundamental frequency and period of the fixed supported system are equal to 29.22 rad/second and 0.21 seconds, respectively. The effective mass as a percentage of the total mass is equal to 90.23%. The fundamental frequency and period of the system with flexibility at its base, which represents a seismically isolated structure, are equal to 4.97 rad/second and 1.26 seconds, respectively. The effective mass as a percentage of the total mass is equal to 99.98%.

Figure 3.1 shows the relative displacement histories of the fixed supported system, and Figure 3.2 shows the relative displacement histories of the same system when flexibility is inserted at its base.

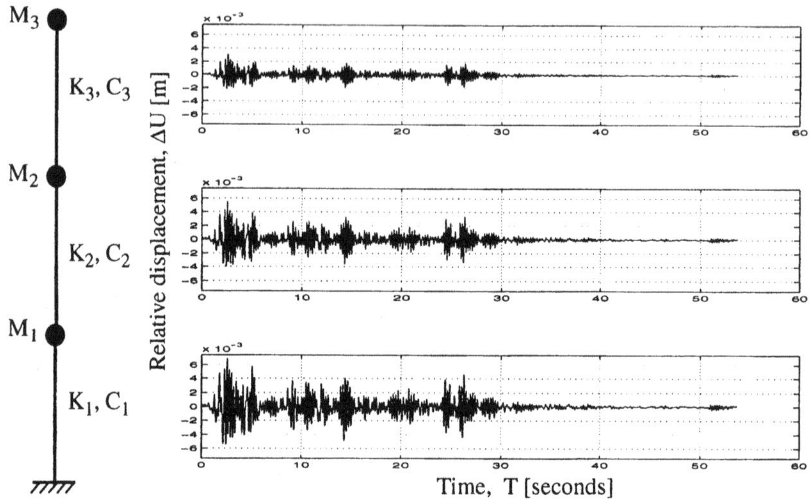

Figure 3.1: Inter-story deflections of a fixed supported system.

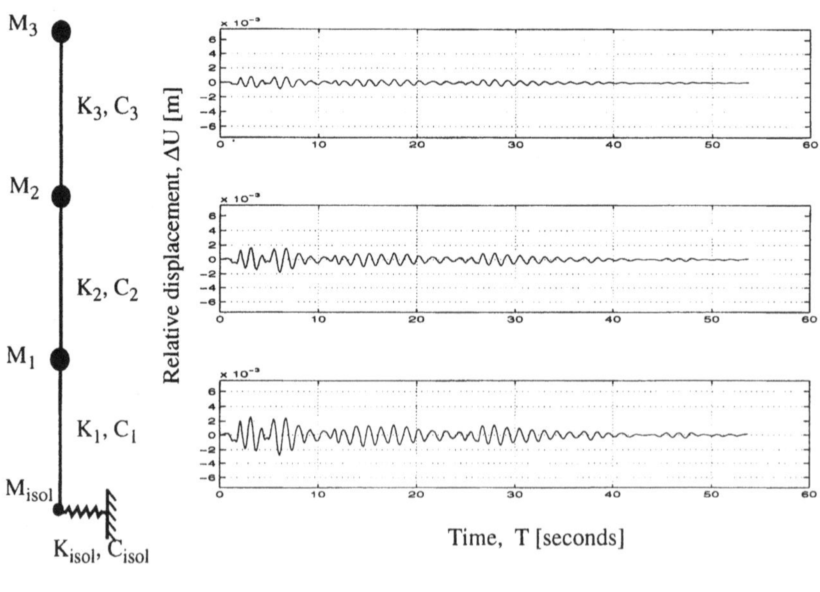

Figure 3.2: Inter-story deflections of a system mounted on a flexible spring.

Figures 3.3 and 3.4 show the shear force histories of the system under fixed-supported and seismically isolated boundary conditions, respectively. The reduction to the shear forces, when flexibility is inserted at the base of the system, is rather significant.

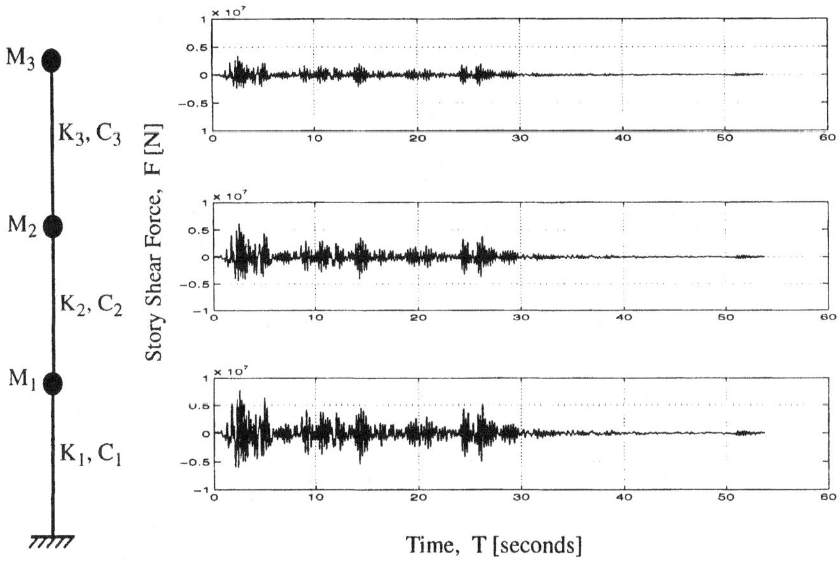

Figure 3.3: Story shear forces of a fixed supported system.

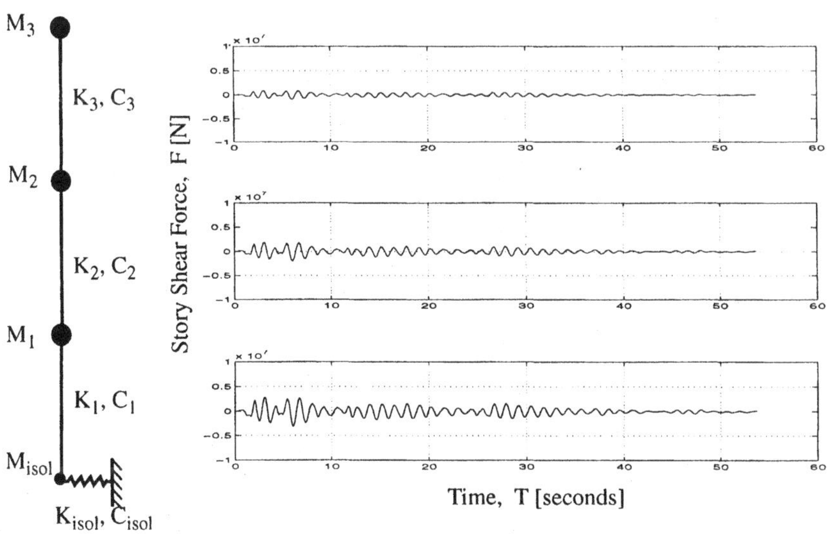

Figure 3.4: Story shear forces of a system mounted on a flexible spring.

Most deformation in a seismically isolated structure occurs at the isolation level and is due mostly to the contribution of the fundamental (i.e. first) mode in which the superstructure remains essentially rigid. The fundamental mode involves most of the energy transmitted to the system and it has the highest participation in the structural response. The deformation of the superstructure is due to the higher modes, which, however, are orthogonal to the first mode. Therefore, the first mode does not allow the seismic energy corresponding to these higher modes to be transmitted to the superstructure, substantially reducing the superstructure deformations.

From the structural characteristics and the response of both models it is evident that when a conventional earthquake-resistant system is used, the seismic forces, relative displacements and accelerations are larger than those occurring when a seismic isolation system is used.

3.5 Constraints in the Implementation

The main constraint in the practical implementation of seismic isolation is the large relative displacements between the superstructure and the supporting ground at the isolation level, which must be ensured for the whole life of the structure. Large relative displacement at the isolation level limits the extent of isolation. It is indicated by the shift of the fundamental period of the isolated structure with respect to that of the fixed supported superstructure. However, this displacement may be reduced by providing an energy dissipation mechanism. Among several such mechanisms the ones most commonly used are based on the plastic deformation of a metal, such as lead or steel, or on the inherent damping properties of some types of rubber.

Another constraint is the possibility of uplift of the isolators. During severe earthquakes, the lateral seismic forces and resulting moments may cause axial loads larger than the gravity loads. In such cases, tensile stresses may be applied to the elastomeric bearings which are not designed to carry these stresses, and the bearings' performance may thus be impoverished. Connection details and proper configuration of the layout of the structure are necessary in order to avoid excessive amounts of uplift forces. The possibility of uplift of the isolators increases with height and particularly with the aspect ratio (height-to-width) of the structure due to the increased overturning moments. This makes the use of seismic isolation inappropriate in high-rise buildings.

Also, isolation of a very flexible structure may be inefficient, since the fundamental frequency of such a structure is usually already outside the dangerous-for-resonance range. Thus, the main feature of seismic isolation, the avoidance of resonance, has no significant effect on flexible structures. In addition, high-rise buildings are more vulnerable to wind loads than to earthquake loads, and seismic isolation is not an appropriate way of handling these loads. Seismic loads do not usually dominate the design of high-rise buildings. Some devices have been developed to provide energy dissipation without the incorporation of isolation, and these are more suitable for high-rise buildings. Active control, either in the form of tuned mass dampers or closed-loop actuators, may be used to mitigate wind excitations.

3.6 Limitations

Although seismic isolation is a very promising design method for dealing with earthquake loads, it can not be used for all structures and at all sites. Some examples have been given in the previous section. A feasibility study must therefore be carried out early in the design phase, to determine the plausibility of this design approach. This study must be based on a number of factors, which are presented next.

3.6.1 Superstructure characteristics

As previously mentioned, seismic isolation is generally suitable for low- to medium-rise buildings (usually up to 10-12 stories) which have their fundamental frequency in the range of the usual dominant frequencies of earthquakes. Superstructure characteristics, such as height, width, aspect ratio and stiffness, are related to the applicability and effectiveness of seismic isolation. Most buildings, worldwide, have the above characteristics, which makes them potential candidates for seismic isolation whenever they happen to be located in a seismic-prone region with a high probability of strong earthquake occurrence.

3.6.2 Seismic hazard and site characteristics

The seismicity of a particular region must be considered, in order to determine the necessity of seismically isolating a structure in that region. Foundation soil characteristics, such as the site period T_s and the expected dominant frequencies of earthquakes in a particular area, are factors which must be considered in a feasibility study. In general, the stiffer the foundation soil, the more effective is seismic isolation. In cases where site characteristics (e.g. soft soils) and information from previous earthquake records indicate low dominant site frequencies, i.e. high T_s, seismic isolation must be avoided. The site period T_s, considering geotechnical characteristics, corresponds to the fundamental vibrational mode of the geotechnical soil profile.

Deep deposits of soft soils usually amplify the low-frequency motions, eliminate the high-frequency motion, and impose low-frequency ground motion on the structure. In these cases, the foundation soil may be treated as a physical partial seismic isolation and the structure must be as rigid as possible. In addition, the site-expected earthquake predominant period tends to be lengthened with distance from the epicenter, due to high-frequency component attenuation. A flexible structure is more vulnerable to low-frequency motion due to resonance.

Finally, another issue which must be considered is the proximity to active faults due to possible long period components of the expected seismic excitation. These are near-fault effects caused when a rupture propagates towards the region with fault rupture propagation velocity equal to the seismic wave velocity, resulting in long unidirectional pulses. The latter may lead to higher displacements than those expected.

3.6.3 Surrounding structures

All adjacent structures or facilities which may impose restrictions on the seismic isolation system must be taken into account, especially in order to estimate the maximum allowable displacement. A clearance around the building, of the order of 10 to 20 cm, must be ensured for its whole life, to accommodate large relative displacements between the superstructure and the ground foundation. Special care must be taken in heavily constructed areas to avoid collision of the structure with surrounding ones.

Note that there are also cases where the location of adjacent structures does not allow the use of seismic isolation. Finally, the surrounding ground may also impose restrictions, and special design details may be needed to enable the use of seismic isolation. For example, a seismically isolated structure located on the slope of a hill must be protected by special design and construction details from the possible application of soil thrust pressure upon it, which would be detrimental to the seismic isolation system.

3.6.4 Minor lateral loads

Wind loads and other lateral loads must be limited, i.e. usually less than 10%; otherwise seismic isolation is inappropriate. It must be ensured that the seismically isolated structure performs well under these minor loads. For bridge structures, loads due to braking or centrifugal forces should be taken into account as well.

3.7 Factors Which Enable the Use of Seismic Isolation

Several factors favor the development and practical application of seismic isolation. First of all, there are increased requirements for the performance of structures in severe earthquakes which are not met by current conventional design philosophy. These are crucial to structures containing sensitive and expensive equipment, vulnerable even to microtremors.

Second, the advance of computer technology and modern structural analysis methods enables the development of reliable software to simulate the response of structures. The development of seismic engineering and earthquake engineering to a level where reliable predictions can be made for expected earthquakes, is another important factor. Today a dynamic time history analysis is easily implemented on a personal computer.

In addition, the construction of shaking tables which can simulate actual earthquake excitations makes possible the experimental validation of the behavior of seismic isolation systems. The study of other minor loads such as wind loads, and the reliable quantification of their expected intensities and frequency of occurrence, also enables the use of seismic isolation.

Finally, development, manufacture and extensive research in the area of structural materials enable the reliable use of modern materials for seismic isolation devices. The development of devices which dissipate energy and provide a restoring force to avoid permanent displacements also allows the practical implementation of the seismic isolation concept.

3.8 Economics of Seismic Isolation

In terms of cost, although seismic isolation usually requires an additional initial cost, it ensures long term savings due to better seismic performance and damage reduction. In particular, an additional initial cost of the order of 1% to 5% of the structural cost may be required in order to incorporate seismic isolation into a new building (Mayes et al. [71]). However, the isolated structure is designed to respond to a strong earthquake without suffering the damage that would be sustained by a conventionally designed fixed-supported structure. When the method is used for seismic upgrading of an existing building, savings may result according to the level of the structure's seismic performance capacity, the desired seismic upgrade, and the already present constraints. In bridge construction or seismic upgrading, seismic isolation usually results in cost savings, especially when high-performance requirements are considered in the design. Most existing bridges may be seismically isolated by simple substitution of the existing thermal expansion bearings, which in many cases have to be replaced anyway due to deterioration, with isolation bearings, and provision of the required gap for the relative displacements of the superstructure.

It is difficult to compare seismic isolation with conventional design in economic terms, due to their different performances, the uncertainties associated with their response, and the exact level of damage each method allows. For a specific earthquake, and with acceptance of the same damage, seismic isolation typically reduces the initial construction cost. An example which supports this argument is that of nuclear plants, for which the earthquake performance requirements are specifically defined and thus their design implies equal performance. A feasibility study for a nuclear power plant (Burns and Roe [19]), showed that incorporation of seismic isolation resulted in a cost savings of $34 million for a project of $1,760 million, i.e. the savings are of the order of 2% of the initial structural cost. A detailed comparison between the cost of seismic isolation and that of conventional earthquake resistance design was made for the Fire Command and Control Facility of Los Angeles County in the United States. An essential design requirement was to ensure that the building would remain functional under the maximum credible earthquake. It was found that seismic isolation would result in approximately 6% savings on the conventional fixed supported design. The significant savings resulted from the reduced mechanical and electrical equipment requirements, because of the smaller imposed seismic loads. In contrast, more stringent equipment and non-structural requirements had to be adopted for the conventionally designed building to reduce non-structural damage, taking the larger seismic loads into consideration (Anderson [5]).

In most cases, the seismic isolation option is compared to conventional options but at different performance levels. The aim of seismic isolation is to give better performance in coping with earthquake excitations, avoiding the damage which results when the structure is fixed supported, rather than to reduce the initial construction cost for thc samc lcvcl of pcrformancc. The goal of seismic isolation, to avoid damage, cannot be achieved with a conventional design of reasonable cost.

There are, however, several factors which may cause an increase in the cost if seismic isolation is used. First, it is the cost of the isolation system itself; a typical lead rubber bearing costs between $5,000 and $25,000. Also, additional members or structural changes may be required in order to insert the isolators; this also implies an additional cost. For example, an additional slab at the isolation level is necessary for uniform deformation and stress distribution in the isolators. Cost of modifications, such as shear walls around the structure to accommodate the large relative displacement of the isolation system, must also be taken into account, as well as the loss of potentially usable area which is required for the seismic gap. This loss may be significant in densely populated areas where land is expensive. Finally, seismically isolated structures require rigorous maintenance and frequent inspections to ensure their adequate performance during the life span of the structure, which results in additional cost. However, the idea of having regular inspections is gaining ground even for conventionally designed structures, especially after some recent disastrous earthquakes.

There are also several factors which may lead to savings when seismic isolation is used. The superstructure construction cost may be lower due to lower seismic forces and reduced seismic capacity demand, which is more evident in seismic retrofitting of existing structures. The reduced seismic loads allow more deliberate design of the superstructure. Thus, construction methods which are not suitable in high seismicity areas may then be used for seismically isolated structures. Such methods include prestressed and precast concrete construction which do not provide sufficient ductility capacity. In addition, considering the substantially smaller inter-story drifts, cheaper non-structural parts, such as cladding, may be used. Long-term savings also result from the reduction in repair costs for structural and non-structural damage, due to the substantially decreased inter-story displacements, and from the increased building contents protection due to reduced floor accelerations. The avoidance of disruption of the structure's operation, due to the better performance of seismic isolation, also results in savings. In general, repairs are very time consuming and the loss of functionality of some structures may be more disastrous than the actual damage.

The use of an effective earthquake-resistant design such as seismic isolation always results in benefits from increased safety and reduced injuries, deaths and related lawsuits. However, savings from insurance costs cannot currently be estimated since most insurance companies do not provide special premiums for seismically isolated structures. Such premiums can be more readily provided, however, for structures with higher reliability than for conventionally designed ones. Insurance companies define such high deductibles and premiums and it may be more beneficial to incorporate seismic isolation to reduce damage below the deductibles, and thus counterbalance the additional initial cost with the reduced amount paid for annual insurance premiums. Some insurance companies take into account the use of seismic isolation, reducing the annual insurance premium for a seismically isolated building and, therefore, amortizing the additional initial cost for the seismic isolation.

3.9 Reasons for the Lack of Extensive Use of Seismic Isolation

The extent of use of seismic isolation is affected by several factors. Clients are not always aware of the objectives, limitations and undesirable consequences of conventional earthquake resistance design. They are often misled to believe that a conventional fixed supported building, designed according to current earthquake resistance codes is absolutely earthquake proof, for earthquakes of any magnitude. In general, people learn of the actual purpose of the current earthquake design code only after a severe earthquake with disastrous consequences. Hence, they rely on the current conventional approach and are not motivated to use a more controversial and yet promising approach, especially when its incorporation requires an additional cost.

Indeed, seismic isolation, in general, requires an additional initial cost and provides savings only in the long run, when different performance requirements are used. It is very difficult to convince people, who are usually interested in immediate savings, to invest in a design which will provide only long-term savings resulting from reduced damage during a possible future earthquake. As long as decisions on the design and construction of a structure are dominated by its initial cost, it is difficult for seismic isolation to be widely used. When more strict requirements for the seismic response and the allowable level of damage are adopted, seismic isolation will be compared to the currently-used design method under approximately the same response requirements, and may be shown to ensure savings even in the initial cost.

Information on seismic isolation available to the public is limited. In addition, the method is not introduced as a topic of study in most universities. Therefore, structural engineers cannot consider it as an alternative earthquake design method. In only a few universities is this method taught or even mentioned to students. Therefore, its practical implementation, which requires exposure of structural engineers to seismic isolation, is limited. Also, design firms involved in seismic isolation projects release only limited pertinent information to the civil engineering community, due to market competitiveness. I believe, however, that the design aids, procedures and any other pertinent information must be published and distributed not only to practicing structural engineers but also to students. The release of information related to seismic isolation and data from manufacturing, design and construction companies, is essential to increasing the spread of knowledge of the method.

Seismic isolation is an innovative and unconventional design idea, and engineers are usually conservative in implementing such ideas. Thus, in the same way as it took a few decades for the use of prestressed concrete in construction to be established, it may take time for seismic isolation to be understood and established. Considering that seismic isolation has been used in many projects during the last 30 years, the time for its general acceptance has possibly come. For the practical implementation of seismic isolation, several people with different interests and backgrounds must work together or independently. In this way, large companies which specialize in seismic isolation should have a staff of engineers and scientists, or even different divisions, which will take care of the different issues involved in the implementation of the method.

Unfortunately, most isolation systems are currently patented and may therefore be used only by the few companies which have the rights on these systems. Architectural and construction technology has been developed and advanced for thousands of years and therefore belongs to society. Constraints on the free use of seismic isolation systems can only cause more casualties and extensive damage, a fact that the companies which have the systems' exclusive rights do not consider, in view of the large profits they make.

Finally, economic incentives must be offered to motivate the wider use of seismic isolation. States should realize potential savings from a reduction in damage during future strong earthquakes and from a reduction in the reimbursements and support that must be provided to victims. Wider use of seismic isolation, supported by the state, will result in lower costs to seismically isolate a structure, since the companies in this industry sector can profit from increased sales rather than from very high prices, which is their current practice. Wider use of seismic isolation may create further cost savings since less conservative codes can be applied as soon as the method matures and sufficient experience is achieved. Once the superior behavior of seismically isolated structures has been observed and the relevant data collected, design codes for seismic isolation may reduce the requirements for the superstructures, i.e. above the isolation level, of seismically isolated structures. This will result in cost savings during construction, making seismic isolation an attractive approach even in economic terms.

Chapter 4

Seismic Isolation Devices

4.1 Types of Seismic Isolation Systems

Various seismic isolation systems have been developed and several have also been applied in practise. Seismic isolation systems used in the United States include ones that consist of elastomeric bearings and sliding systems. Figure 4.1 shows the percentage use of these systems in the United States as of January, 1994.

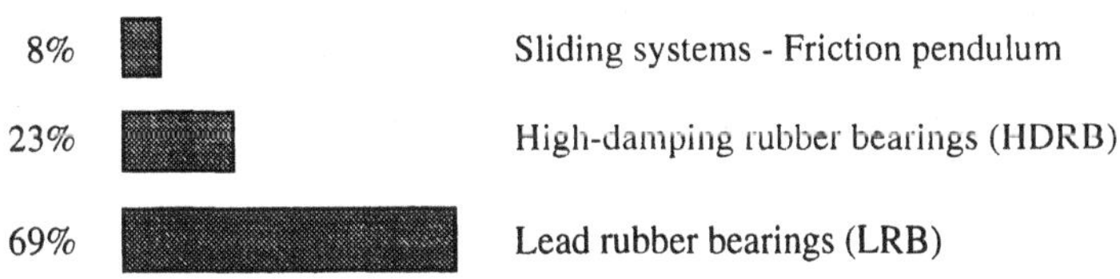

Figure 4.1: Seismic isolation systems used in the United States as of January 1994.

The above systems combine simplicity, reliability and economy. In addition, there is substantial experience available, at least with elastomeric bearings since they have already been used extensively in bridge construction to accommodate deformations due to thermal changes. Several other systems have been developed, but not so successfully.

4.2 Sliding Isolation Systems

Sliding systems, which are based on friction, allow transmission of shear forces up to a particular level, beyond which sliding occurs and transmission is prevented. Such systems may be designed to allow a very low base shear force using a low coefficient of friction. The most commonly used sliding systems have sliding surfaces of stainless steel and teflon. The magnitude of the shear that can be transmitted to the superstructure during a strong earthquake is independent of the severity of the earthquake since the transmitted force is based on the friction coefficient. Thus, sliding systems are very efficient in mitigating the effects of severe earthquake excitations on structural systems. In addition, they are relatively cheap and of compact size and thus appropriate to be used in retrofits of existing structures.

However, many sliding isolation systems may not return to their initial configuration after the excitation, due to absence of a restoring force. Permanent offset of the isolated structure from its original location may result, decreasing the amount of displacement that can be accommodated during a following earthquake.

A restoring force can be provided by elastomeric bearings which can be combined with sliding systems. Such a combination may be able to utilize the advantages of both isolation systems. Besides providing a restoring force, the use of elastomeric bearings may also provide a resistance against extremely strong earthquakes by the stiffening of the rubber under very large strains.

Another way to provide a restoring force is used in the friction pendulum system (FPS), which makes use of spherically shaped sliding bearings (Figure 4.2). The system consists of a spherical stainless steel surface, whose geometry generates the self-centering action, and a slider.

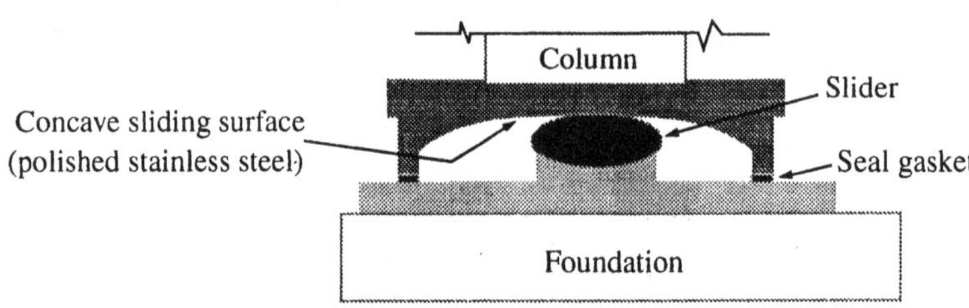

Figure 4.2: Sliding isolation system: friction pendulum system (FPS).

During an earthquake the structure can slide on the isolator. The word pendulum is used because the system resembles a pendulum's motion when considering the friction due to the relative movement of the one part of the isolator with respect to the other. The restoring force is generated as the structure moves relative to the curved spherical surface of the isolators, sliding horizontally, and rises, moving upwards due to the curvature. The force that is required to move the structure limits the lateral forces that can be exerted on it, i.e. the base shear. The lateral force that is required to slide the structure laterally depends on the radius of curvature of the isolator and the vertical load. The geometry, essentially the radius of curvature, is properly selected in order to provide the desirable effective stiffness of the isolator and, consequently, the fundamental frequency of the isolated structure. The energy dissipation is generated by the friction between the slider and the spherical surface.

However, the significant difference between the initial, i.e. prior to sliding, and the post-yield stiffness may cause high-frequency effects. The sudden changes of stiffness due to the sudden slipping and sticking effects may generate high-frequency vibrations of the structure. These may in turn increase the floor accelerations and consequently the possibility of damage to the structure's contents. Therefore, sliding systems are not suitable for structures with content protection requirements.

Another disadvantage is that the coefficient of friction is not constant; it may vary with temperature and time due to material deterioration. This may cause an increase in the actually inserted seismic forces, resulting in severe damage if lower seismic forces have been considered in the design. The combination of sliding systems with elastomeric bearings results in hybrid systems which may have the advantages of both individual systems, but also some of their disadvantages and limitations, to a lower degree however.

4.3 Elastomeric Isolation Systems

The most popular seismic isolation systems use elastomeric bearings (Figure 4.3) which consist of thin rubber sheets bonded onto thin steel plates and combined with an energy dissipation mechanism. The rubber sheets are vulcanized and bonded to the thin steel plates under pressure and heat.

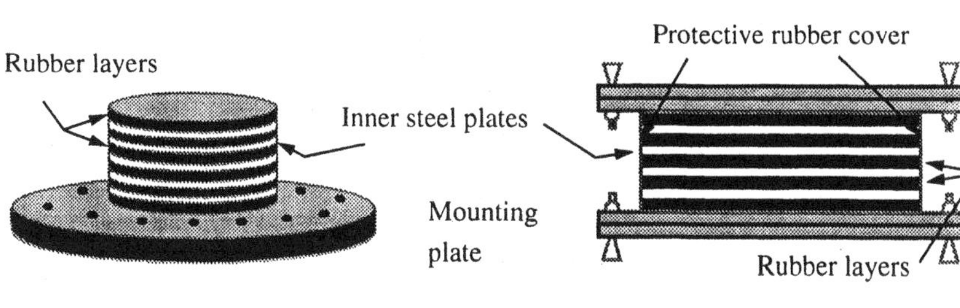

Figure 4.3: Elastomeric bearing.

The inner thin steel plates provide the vertical load capacity and stiffness, and prevent lateral bulging of the rubber. In particular, the steel plates laterally constrain the rubber sheets as vertical load is applied to the elastomeric bearing, providing the vertical stiffness. Horizontal flexibility is provided by the shearing deformability of the rubber sheets, which are not restrained from deform in that direction by the steel plates. Thick mounting steel plates are bonded to the bottom and top surfaces allowing the isolator to be firmly connected to the foundation below and the superstructure above.

The energy dissipation mechanism is based either on the plastic deformation of a metal or on the inherent damping properties of the rubber. In the first case, either lead plugs are inserted in the elastomeric bearings, or auxiliary dampers based on plastic deformations of lead or steel are used.

Elastomeric bearings have minimum maintenance requirements, since they have no moving parts or components exposed to corrosion conditions. However, the cost of elastomeric bearings is still relatively high, prohibiting their wide use in general housing and commercial buildings. The cost of elastomeric bearings could be reduced by an increase of the volume of sales, which will happen when seismic isolation is more widely used. The production cost of elastomeric bearings could also be reduced by automating the manufacturing process. The high cost is due to the special manufacturing requirements and the labor cost. The steel plates need to be cut to the exact sizes, sand-blasted and chemically cleaned. Then they have to be coated with a bonding compound, and the rubber sheets laid on them in a mold for vulcanization bonding. The interleaved steel plates and rubber sheets are headed under pressure for several hours. Finally, cover rubber is used to wrap the elastomeric bearing as an additional measure to protect the steel plates from corrosion. In addition, special fire-resistant coverings can be used to protect the isolators from fire hazards.

Lead rubber bearings (LRBs) and high-damping rubber bearings (HDRBs) are most useful in seismic isolation since they provide the following in a single unit:

• Vertical support due to the high vertical stiffness, which is usually several hundred times the horizontal stiffness. Sufficient vertical stiffness is necessary to avoid rocking of the structure.

• Horizontal flexibility which shifts the fundamental frequency of the structure out of the dangerous-for-resonance frequency range.

• An energy dissipation mechanism, either via the plastic deformation of the lead plug or through the inherent damping properties of high damping rubber.

In addition, a lead rubber bearing provides initial rigidity under service lateral loads, such as wind loads, due to the presence of the lead plug and its high stiffness prior to yielding. In that case, however, the energy dissipation mechanism is activated only once the lead plug has yielded. High-damping rubber bearings provide an energy dissipation mechanism on a continuous base, but not the above-mentioned initial rigidity. Lead rubber bearings, furthermore, provide the greater restoring effect which is necessary to re-center the isolators at their original locations after the end of the earthquake.

Finally, there are also some systems that use natural rubber bearings (NRBs) with an additional steel or lead damper; in this case energy dissipation results from the plastic deformations of the damper.

4.3.1 Lead rubber bearings (LRBs)

This type of elastomeric bearing consists of thin layers of low-damping natural rubber sandwiched between steel plates, and a lead cylinder plug firmly fitted in a hole at its center to deform in pure shear (Figure 4.4 and Figure 4.5). The inner thin steel plates confine the lead core.

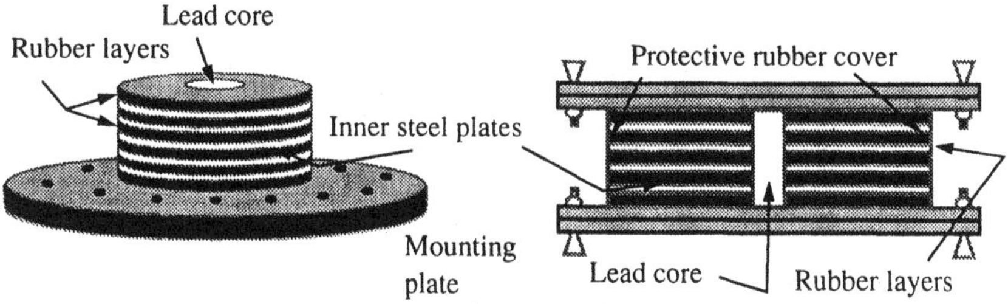

Figure 4.4: Lead rubber bearing.

The following photographs (Figure 4.5) show a cross-section of a lead rubber bearing and a lead rubber bearing installed in a seismically isolated building.

Figure 4.5: Lead rubber bearing.

Low-damping natural rubber is, usually, used in lead rubber bearings. The horizontal flexibility is achieved by the very low shear modulus of the rubber, which exhibits a linear behavior for strains up to 100-250%. Tests have shown that its mechanical properties are not influenced by temperature, aging, and rate or history of loading.

Lead is a crystalline material which changes its structure temporarily, under deformations beyond its yield point, and regains its original structure and elastic properties as soon as the deformation is removed by the restoring force in the rubber. As the lead is forced by the steel plates to deform plastically in shear, once it has exceeded its yield stress, it dissipates energy hysteretically. Lead has an elastic plastic behavior with a yield stress σ_y of approximately 10.5 MPa. The strength of lead is much less than the theoretically computed one from the inter-atomic distance-force curve, due to the presence of defects termed dislocations. These allow the bonds to break, one at a time, causing the material to yield and deform plastically for relatively low stress. Note that lead has good fatigue properties for subsequent cycles of loading beyond its yield point.

Hysteretic energy dissipation is equal to the area offset of the loading and unloading curves under cycling loads. At each cycle this area represents the energy dissipated in the form of heat. It is actually due to work done during loading which did not completely recover during unloading, but was converted from kinetic to thermal energy and dissipated. However, the non-linearities due to the lead plug deformations, although they provide damping, may also induce responses in the higher modes of the structure. In particular, the continuous sudden changes of stiffness of LRBs may excite higher modal response effects, lowering the anticipated reductions of the accelerations of the structural masses.

The behavior of the LRB is characterized by a high initial, i.e. prior to yielding, stiffness K_{el}, due to the presence of the lead plug K_l, and by a low post-yield stiffness K_{pl} which is equal to the shear stiffness K_r of the rubber (Figure 4.6). The force required for a particular displacement, when the yield force of lead has been exceeded for the first time, is equal to the sum of the two individual forces, since they act as springs in parallel. The force prior to yielding is equal to $F = \sigma_y A + K_r u = (K_{el} + K_{pl}) u$. After yielding, that force becomes equal to $F = K_r u = K_{pl} u$. For reversed cyclic loads the exact history of the loading is required to obtain the force F.

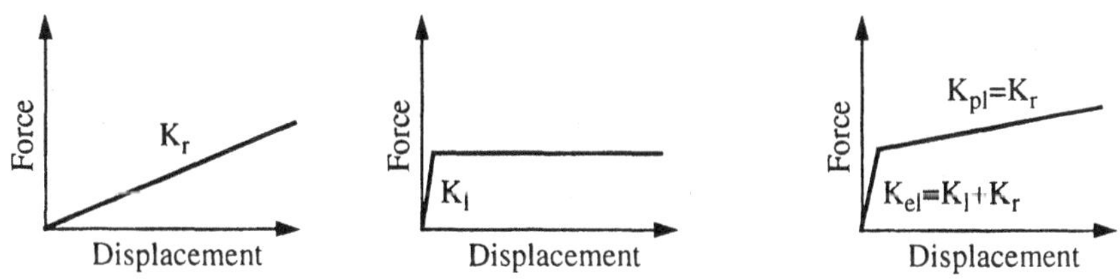

(a) Rubber (b) Lead (c) Lead-Rubber

Figure 4.6: Mechanical behavior of LRBs.

Usually, the behavior of the LRB is modeled by an equivalent linear viscously damped system with an effective stiffness K_{eff} and an equivalent viscous damping ratio ξ_{eff}, as will be described in Chapter 7. However, this is not always a good approximation, since the equivalent viscous damping varies considerably according to both the specific characteristics and the intensity of an earthquake, as well as the characteristics of the isolation system and the superstructure. Lead rubber bearings provide initial rigidity, due to the high elastic stiffness of lead, which is essential for minor lateral loads. The presence of the lead plugs, the variable horizontal effective stiffness, and the variable effective fundamental period of the structure, enable the structure to vibrate outside the dangerous-for-resonance frequency range as the response increases.

The initial stiffness of the LRB is very high but as the intensity of the excitation increases, the stiffness is reduced and the isolation system becomes more effective. The initial elastic stiffness K_{el} is approximately ten times the post-yield stiffness K_{pl}. The horizontal stiffness of the LRB decreases as shear strain increases; beyond 250-300% strain it increases again due to hardening effects (Robinson [97]). Strain hardening effects of elastomeric bearings under excessive deformations are useful as a safety valve against severe earthquake excitations. The breaking point occurs at shear strain levels up to 500% and is found to be independent of compressive or tensile stresses and the presence of the lead plug. The shear strain may also be considered as independent of the vertical load. However, the vertical stiffness depends on shear deformations.

The elastomeric bearings can easily undergo large displacements withstanding very large deformations (Figure 4.7). Failure of a bearing occurs mainly due to the formation and growth of flaws in the elastomer. The extent of rupture depends on the compounding and stiffness of the material. Careful manufacturing and quality assurance control can prevent flaws in the elastomer that may lead to its failure under extreme loading conditions.

Figure 4.7: Deformed elastomeric bearing.

It has previously been mentioned that lead rubber bearings do not provide an energy dissipation mechanism under service loads, i.e. prior to the yielding of the lead plugs. This makes them ineffective for dissipating energy against microtremors or minor lateral loads which may affect structures with sensitive equipment. Auxiliary dampers may therefore be incorporated to dissipate energy prior to yielding. Also, the actually bilinear behavior of the LRB results in a non-constant equivalent damping ratio which varies with both the excitation characteristics (intensity, dominant frequencies, etc.) and the isolated structure's characteristics.

Finally, the strain energy stored during the earthquake in the lead plug and the elastomer provides the necessary restoring force that re-centers the structure at its initial configuration after an earthquake. The presence of this restoring force has been validated in practise whenever seismically isolated structures mounted on LRBs have experienced earthquake excitations. In addition, shake table tests have demonstrated the ability of the LRBs to restore a structure to its original configuration. All numerical and experimental tests have shown that the elastomeric bearings had returned to their original undeformed position within 2.5 centimeters.

4.3.2 High-damping rubber bearings (HDRBs)

This type of bearing consists of thin layers of high damping rubber sandwiched between steel plates. The same manufacturing methods for vulcanization and bonding that are used for LRBs are also used to construct HDRBs. The only difference is the composition of the rubber compound, which provides increased damping.

High-damping rubber is actually a filled rubber compound with inherent damping properties due to the addition of special fillers, such as carbon and resins. The addition of fillers increases the inherent damping properties of rubber without affecting its mechanical properties. When shear stresses are applied to high-damping rubber, a sliding of molecules generates frictional heat which is a mechanism of energy dissipation. In unfilled natural rubber, used for LRBs, frictional heat is negligible because the molecular attraction in physical cross links is very weak. The energy dissipation mechanism of an HDRB is available for both small and large strains, is constant and is characterized by smooth elliptical hysteresis loops. Experimental studies of high-damping rubber bearings verified the anticipated energy dissipation capacity, which is, typically, equivalent to about 15% damping ratio of equivalent linear elastic models.

However, an HDRB may not provide the necessary initial rigidity under service loads and minor lateral loads, although some initial rigidity is provided by high-damping rubber compounds which exhibit higher stiffness under small strains. A structure isolated with HDRBs essentially has a constant, large fundamental period due to the flexibility of the isolation system, which makes the structure vulnerable to wind action with dominant frequencies close to the fundamental frequency. In addition, the damping and mechanical properties of the HDRB appear to be temperature dependent while the hysteretic energy dissipation mechanism of the LRB is not. HDRBs are not as widely used in seismic isolation as LRBs.

The shear modulus of high-damping rubber is usually between 0.3-1.5 MPa. Certain high-damping natural rubber compounds exhibit mechanical properties and behavior that make them suitable for seismic isolation. In particular, the initial stiffness of the high-damping rubber is high at small strains and decreases substantially with increasing strains reaching a minimum value approximately equal to 20% of the initial stiffness at a shear strain equal to about 0.5. The stiffness increases again for strains larger than 100% due to strain hardening effects. This mechanical property provides some high initial stiffness, which is useful for wind or low-intensity earthquake loadings. In addition, the high stiffness under large strains provides a safety valve for cases of extreme earthquake excitations in which displacements may exceed the maximum anticipated values.

4.3.3 Hybrid type: lead high-damping rubber bearing (LHDRBs)

A hybrid type of lead high-damping rubber bearing (LHDRB) may consist of layers of high-damping rubber sandwiched between steel plates and a smaller diameter lead cylinder plug firmly fitted in a hole at its center. This hybrid bearing may have the advantages of both isolation systems discussed above. The LHDRB has both an initial rigidity, due to the presence of the lead plug, and a continuous energy dissipation mechanism, due to the damping properties of the high-damping rubber. Therefore, the isolation system is expected to perform well in both weak and extreme earthquakes as well as under minor lateral loads. In addition, the properties of this bearing will be less dependent on temperature changes and shear strains than those of an HDRB. Also, the use of LHDRBs allows the reduction of the initial stiffness of the isolation system which is responsible for higher frequency effects. A final advantage is that the effectiveness of such a device and its compact size make it suitable for cases where installation space is limited. LHDRBs offer a practical and cost-effective trade-off between the advantages and limitations of LRBs and HDRBs.

When practically possible, it may be more effective to place the high-damping rubber bearings with lead plugs at the perimeter of the building, preferably under the columns situated as far away as possible from the center of stiffness and mass of the isolated structure. High-damping rubber bearings with no lead plugs may be used under the internal columns. This configuration will allow lower prior-to-yielding stiffness of the lead plugs and, consequently, a smoother change of the stiffness during yielding and reverse loading. The high initial stiffness and its sudden changes are responsible for higher mode effects and acceleration increases, which may be avoided by reducing the initial stiffness. In addition, lower initial stiffness will provide a higher degree of isolation at the prior-to-yielding stage. Placing the bearings with the lead plugs under the external columns away from the center of stiffness will provide higher resistance against torsion, due to the larger distance between the points of application of forces and the center of stiffness of the isolation system. Note that this configuration may be used only when there is a rigid diaphragm at the isolation level to redistribute the inertia forces to the LHDRBs.

An effective isolation system must have both viscous and hysteretic damping and must ensure a continuous energy dissipation mechanism. The viscous damping, which is velocity dependent, will ensure a continuous energy dissipation mechanism for both severe earthquakes and microtremors. Viscous damping may be provided by actual viscous dampers or by rubber with inherent damping properties. The latter does not actually provide such damping, which may, however, be assumed due to the rubber's smooth elliptical hysteresis loops during cyclic loading. The optimum viscous damping ratio lies within 20 to 30%; higher values lead to increases in floor accelerations.

Viscously damped systems are not so effective for severe excitations and do not, in general, provide high initial stiffness. As already mentioned, hysteretic damping is usually provided by the plastic deformation of a metal, typically lead or steel, and is very effective for strong motions, substantially reducing the response, especially the relative displacement at the isolation level. It provides high initial stiffness, required for service lateral loads, but not a continuous dissipation mechanism. The sudden stiffness changes may increase the floor accelerations. Therefore, a compromise between these two damping extremes would be optimum in the design for a seismically isolated structure.

4.4 Elastomeric Bearings: Mechanical and Physical Characteristics

4.4.1 Characteristics

Elastomeric bearings must be strong and stiff for vertical loadings and flexible under shear stresses. The properties of the bearings can be controlled by the proper selection of the elastomer compound and the bearing geometry. The bearings must be designed to be sufficiently large to be able to support, with an appropriate safety factor, the maximum expected vertical loads or the service vertical loads considering the corresponding maximum horizontal displacement. In the latter case, vertical forces are transmitted through an active area of the bearing due to deformations. Steel laminates are bonded to the rubber to restrict bulging under compressive stresses. The thickness of the steel laminates t_s and the thickness of the rubber sheets t_r determine the vertical stiffness of the bearing. The vertical load capacity of a bearing increases as the thickness of the rubber sheets is reduced. The total height h_r of the rubber ($h_r = n_r \ t_r$), is determined by the required horizontal flexibility to be inserted, taking into account the mechanical properties of the elastomer.

The diameter of the lead plugs of the LRBs is determined by the required initial rigidity for minor lateral loads. The area of the lead plugs must be such that the corresponding shear forces may be transmitted without yielding. The yield force level is usually defined as a percentage (typically of the order of 5%) of the weight of the structure. The diameter of HDRBs may be chosen in a similar way to transmit minor lateral loads or a fraction of them without yielding. Finally, the location and dimensions of the bearings must be such that torsional vibrations of the structure are avoided.This is easily achieved by forcing the center of stiffness of the isolation system and the center of mass of the superstructure to coincide. When LRBs or LHDRBs are used, special care must be taken to ensure that the above-mentioned centers coincide, both prior to and after yielding of the lead plugs.

It must be ensured that the mechanical and damping properties of the bearings remain constant over the whole life of the structure. In addition, design details, as will be described in Chapter 6, must be used to allow replacement of malfunctioning units. Thick rubber layers encasing the bearings, and antioxidants and antiozonants used in the manufacturing of the elastomer, can minimize the environmental effects. The current condition of natural rubber bearings which have been used in bridges for more than a century shows that their life expectancy is at least 100 years. However, recent testing of elastomeric bearings that were tested 45 years ago has shown some stiffening. The latter finding was probably due to aging mechanisms and such stiffening does not cause a significant increase in the forces that can be transmitted to the superstructure. Finally, in extensive shake-table tests of elastomeric bearings their force-deflection properties remained stable throughout.

The elastomer which is used in the manufacture of the bearings is characterized by its ability to return to its original configuration and dimensions when unloaded. It usually consists of natural rubber (NR-natural polyisoprene), neoprene (CR-chloroprene), EPDM (ethylene propylene diene monomer), or nitrile butadiene rubber (NBR). Natural rubber and neoprene are very well-known materials and are extensively used in common bridge bearings. The chemical composition and the manufacturing process determine the mechanical and physical properties of the elastomer. This has long randomly-oriented molecule chains, and cross-links between the chains introduced during the vulcanization process. The molecule chains are held together by sparse strong covalent cross-links and weak Van der Waals bonds, when below the glass-transition temperature T_g. The latter bonds are due to the non-uniform distribution of electrons; the former are due to sharing of electrons between atoms.

When the elastomer is deformed, the molecules tend to align, thus changing the entropy of the material. The high flexibility of the elastomers is due to this change in the entropy. The restoring force and the shear and Young's moduli are mainly due to the reordering of the chains and their tendency to return to a random configuration. When a load is applied in all three directions, reordering cannot take place and the stiffness is very high; thus the bulk modulus of rubber is much higher than its Young's and shear moduli. Although nonlinear, the behavior of the elastomer is elastic since it returns to its initial configuration, and its properties return to their original values when the applied load is removed. At very small strains, the elastomer is essentially linear while at higher strains the molecule chains change orientation and align in the direction of the loading.

The mechanical properties of the elastomer are time and temperature dependent, but not to levels that may create problems. In general, its stiffness and damping properties decrease as temperature increases. The bearing may stiffen in low temperatures, with the danger of possible crystallization of the elastomer when exposed to very low temperatures for a long time. For some regions where temperatures are very low, thermal stiffening of the elastomer may be a serious problem since it reduces the flexibility of the isolation system and consequently the degree of isolation. In such cases, this problem must definitely be considered and an appropriate elastomer compound must be selected. In particular, neoprene becomes extremely stiff at approximately -40°C, approaching a glass transition condition, while natural rubber approaches a glass transition condition at -55°C (Roeder et al. [98]). At such low temperatures the elastomer bearings may transfer forces which are several times larger than those expected and the assumed isolated structure may actually be fixed supported.

The temperature dependence of the material properties must be taken into account, to avoid the use of incorrect values measured under different conditions from those actually on site. According to experiments (Stanton and Roeder [104]) and (Robinson [97]), deformation capacity and ultimate strength have no significant time dependencies. Measurements of the Young's and shear moduli depend on the method used. Dynamic tests give much higher values, up to twice the statically measured ones. It is therefore more reasonable to use dynamically measured values. The Young's and shear moduli may be determined using a hardness test (Stanton and Roeder [99]).

Elastomers sufficiently resist the action of oil and chemical substances, assuming that only their surfaces are exposed to such effects. Bridge rubber bearings have been used extensively for very long periods with no significant deterioration problems. Natural rubber has been used extensively for automobile engine mounts under long continuous exposure to oils, without significant damage. Ozone concentration for long periods may result in ozone cracking, which occurs mainly when the rubber is in tension, a situation which is not of concern in the case of bearings. Ozone cracking may be reduced or even eliminated via the addition of an antiozonant during the compounding process. A protective end and side rubber cover may be used to protect the bearings from weather, moisture, ozone and other dangerous effects. The rubber cover also increases the bearings' fire resistance which is, however, already satisfactory, according to related experiments. Finally, bonding of the rubber sheets onto the steel plates does not create any problems since the strength of the adhesion between rubber and steel plates is greater than that of the rubber itself.

Elastomeric bearings of various sizes have been designed, manufactured and used for seismic isolation. Bearings that can carry vertical loads ranging from one to ten thousand tons are available. Large elastomeric bearings may weigh over one ton and have a diameter of more than 1.5 meters (Figure 4.8). Bearings whose top plate can displace more than one meter relative to the bottom plate have been manufactured.

Figure 4.8: Large elastomeric bearing (Takenaka Corporation).

4.4.2 Design considerations

Two shape factors, S_1 and S_2, are usually used to describe the geometric characteristics of elastomeric bearings which are related to their mechanical properties. The primary shape factor S_1 is defined as the ratio of the area of the rubber constrained by the steel plates to the free surface area. Appropriate values of the primary shape factor are between 10 and 20 (Kelly [58]). This factor influences the vertical displacement; as S_1 increases, the vertical stiffness increases. For a circular bearing, S_1 is given by

$$S_1 = \frac{Loaded\ area\ of\ rubber\ layer}{Force\ free\ area\ of\ rubber\ layer} = \frac{\pi\ D^2}{4\ \pi\ D\ t_r} = \frac{D}{4\ t_r} \tag{4.1}$$

The secondary shape factor S_2 is defined as the ratio of the diameter of the rubber to the total thickness of the rubber layers

$$S_2 = \frac{D}{n_r\ t_r} \tag{4.2}$$

where D is the diameter of the bearing, t_r the thickness of the rubber sheet, and n_r the number of rubber sheets.

Although the behavior of the elastomer is nonlinear, a linear elastic analysis is useful for the design of elastomeric bearings. The following assumptions are made in such an analysis:

• At any point $\sigma_{xx} = \sigma_{yy} = \sigma_{zz} = p(x, y)$ (Figure 4.9(a)).

• The distribution of pressure $p(x,y)$ is parabolic and the pressure is zero on the bearing faces and circumference (Figure 4.9(b)).

• Points lying on a vertical line prior to loading lie on a parabola after deformation (Figure 4.9(c)).

• Horizontal planes remain horizontal after deformation.

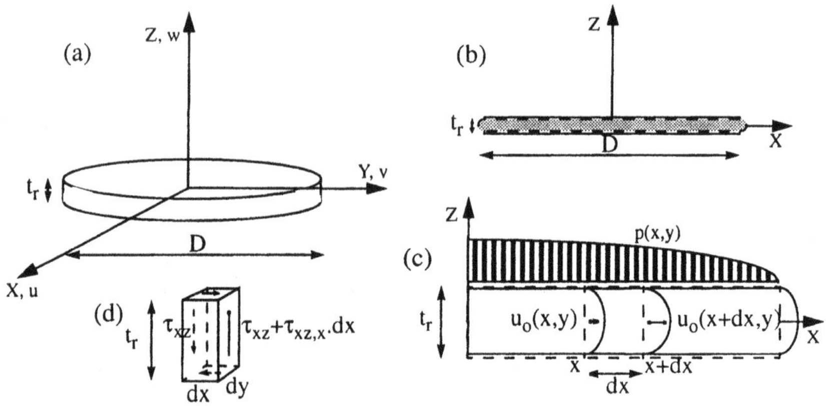

Figure 4.9: Deformation of a rubber sheet.

Using the above assumptions, the displacements are given by

$$u(x, y, z) = u_o(x, y)\left(1 - \left(\frac{2z}{t_r}\right)^2\right) \tag{4.3}$$

$$v(x, y, z) = v_o(x, y)\left(1 - \left(\frac{2z}{t_r}\right)^2\right)$$

$$w(x, y, z) = w(z)$$

The change of volume of an infinitesimal element of dimensions dx, dy and t_r in the X, Y and Z directions, respectively, is

$$\Delta V = \varepsilon_c\, t\, dx\, dy - \frac{\partial}{\partial x}\left(\frac{4}{6}\, t_r\, u_o(x, y)\right) dx\, dy - \frac{\partial}{\partial y}\left(\frac{4}{6}\, t_r\, v_o(x, y)\right) dx\, dy \tag{4.4}$$

where ε_c is the compressive strain. This volumetric change is related to the bulk modulus K of the elastomer through the expression

$$\Delta V = \frac{pV}{K} = \frac{p\ dx\ dy\ t_r}{K}$$

Thus,

$$\frac{p}{K} = \varepsilon_c - \frac{2}{3}\left(\frac{\partial}{\partial x}u_o(x, y) + \frac{\partial}{\partial y}v_o(x, y)\right) \tag{4.5}$$

(Bulk and shear moduli are $K = \dfrac{E}{3(1-2\nu)}$ and $G = \dfrac{E}{2\ (1+\nu)}$, respectively.)

Shear strains are related to shear stresses and deformations by

$$\tau_{xz} = G\gamma_{xz} = G\left(\frac{\partial}{\partial x}w(z) + \frac{\partial}{\partial z}u(x, y, z)\right) = G\frac{\partial}{\partial z}u(x, y, z) \tag{4.6}$$

$$\tau_{yz} = G\gamma_{yz} = G\left(\frac{\partial}{\partial y}w(z) + v\right) = G\frac{\partial}{\partial z}v(x, y, z)$$

Substituting the expressions for displacements, after proper differentiation, we obtain

$$\tau_{xz} = G\frac{\partial}{\partial z}u(x, y, z) = -\frac{8Gzu_o(x, y)}{t_r^2} \tag{4.7}$$

$$\tau_{yz} = G\frac{\partial}{\partial z}v(x, y, z) = -\frac{8Gzv_o(x, y)}{t_r^2}$$

Imposing equilibrium we have

$$\frac{\partial\tau_{xx}}{\partial x} + \frac{\partial\tau_{zx}}{\partial z} = 0 \tag{4.8}$$

$$\frac{\partial\tau_{yy}}{\partial y} + \frac{\partial\tau_{zy}}{\partial z} = 0$$

$$\frac{\partial\tau_{xx}}{\partial x} = -\frac{\partial\tau_{zx}}{\partial z} = \frac{8Gu_o(x, y)}{t_r^2} \tag{4.9}$$

$$\frac{\partial\tau_{yy}}{\partial y} = -\frac{\partial\tau_{zy}}{\partial z} = \frac{8Gv_o(x, y)}{t_r^2}$$

Thus,

$$u_o(x,y) = \frac{t_r^2}{8\,G}\frac{\partial\tau_{xx}}{\partial x} \Rightarrow \frac{\partial}{\partial x}u_o(x,y) = \frac{t_r^2}{8\,G}\frac{\partial^2\tau_{xx}}{\partial x^2} \tag{4.10}$$

$$v_o(x,y) = \frac{t_r^2}{8\,G}\frac{\partial\tau_{yy}}{\partial y} \Rightarrow \frac{\partial}{\partial x}v_o(x,y) = \frac{t_r^2}{8\,G}\frac{\partial^2\tau_{yy}}{\partial y^2}$$

Substituting the above expressions in Equation 4.5, we obtain:

$$\frac{\partial^2\tau_{xx}}{\partial x^2} + \frac{\partial^2\tau_{yy}}{\partial y^2} = \frac{12\,G}{t_r^2}\left(\varepsilon_c - \frac{p}{K}\right) \Leftrightarrow \tag{4.11}$$

$$\frac{\partial^2}{\partial x^2}p(x,y) + \frac{\partial^2}{\partial y^2}p(x,y) = \frac{12\,G}{t_r^2}\left(\varepsilon_c - \frac{p}{K}\right)$$

The resulting partial differential equation relates the vertical strain ε_c to the applied load $p(x,y)$. For circular bearings it is more convenient to use a cylindrical coordinate system (r,θ,z), instead of the cartesian system (x,y,z). In such a case we have:

$$\frac{\partial^2}{\partial r^2}p(r,\theta) + \frac{1}{r}\frac{\partial}{\partial r}p(r,\theta) + \frac{1}{r^2}\frac{\partial^2}{\partial\theta^2}p(r,\theta) = \frac{12\,G}{t_r^2}\left(\varepsilon_c - \frac{p}{K}\right) \tag{4.12}$$

$$fi\,\frac{\partial^2}{\partial r^2}p(r,\theta) + \frac{1}{r}\cdot\frac{\partial}{\partial r}p(r,\theta) = \frac{12\,G}{t^2}\left(\varepsilon_c - \frac{p}{K}\right)$$

The bulk modulus of rubber is relatively very large, and it is therefore reasonable to assume that rubber is incompressible. Consequently, the volumetric change is assumed to be zero. The governing differential equation is reduced to a Poisson equation with boundary condition $p\left(r = R = \frac{D}{2}\right) = 0$, i.e. zero pressure at the circumference of the bearing.

$$\frac{\partial^2}{\partial r^2}p(r,\theta) + \frac{1}{r}\frac{\partial}{\partial r}p(r,\theta) = \frac{12\,G}{t_r^2}\varepsilon_c \tag{4.13}$$

The solution to the above equation is

$$p(r, \theta) = p(r) = \frac{3\ G}{t_r^2}(R^2 - r^2)\ \varepsilon_c \tag{4.14}$$

Integrating the pressure over the bearing area, we obtain the load P as

$$P = \int_{r=0}^{r=R} 2\ \pi\ r\ p(r)\ dr = \frac{3\ G\ \pi\ R^4\ \varepsilon_c}{2\ t_r^2} \tag{4.15}$$

Considering that for a circular bearing $S_1 = \frac{R}{2\ t_r}$ and that $A = \pi\ R^2$ we can write

$$P = 6\ G\ S_1^2\ A\ \varepsilon_c \tag{4.16}$$

We now define the compression modulus E_c as

$$E_c \approx E_c(\gamma) = \frac{P}{A\ \varepsilon_c} = 6\ G\ S_1^2 \tag{4.17}$$

We may obtain an expression for the vertical stiffness $K_v(\gamma)$, neglecting axial deformations, and considering the shear strains of rubber as a function of the primary shape factor:

$$K_v(\gamma) = \frac{E_c(\gamma)\ A}{h_r} = \frac{6\ G\ S_1^2\ A}{h_r} \tag{4.18}$$

Axial deformations may be taken into account assuming that the stiffness from shear and axial strains of rubber act as springs in parallel. Then, the vertical stiffness may be expressed as

$$K_v = K_v(\gamma) + K_v(a) = \frac{6\ G\ S_1^2\ A + E\ A}{h_r} \tag{4.19}$$

Considering that the Poisson ratio ν for rubber is approximately 0.5, the Young's modulus is $E = 3\ G$. Therefore, the vertical stiffness K_v may be expressed as a function of the effective Young's modulus E_c. This takes into account the restraints against lateral displacement imposed by the steel sheets.

$$K_v = \frac{6\ G\ S_1^2\ A + E\ A}{h_r} = \frac{(6\ G\ S_1^2 + 3\ G)\ A}{h_r} \tag{4.20}$$

$$E_c = 6\ G\ S_1^2 + 3\ G = E\ (2\ S_1^2 + 1) \tag{4.21}$$

The volumetric change can also be considered, assuming that the above stiffness, and the stiffness from the volume change of rubber, act as springs in parallel. The following expression for the effective modulus of elasticity E_{cv} in compression considers both shear and axial deformations, as well as the compressibility of rubber.

$$\frac{1}{E_{cv}} = \frac{1}{E_c} + \frac{1}{K} = \frac{1}{6\ G\ S_1^2 + 3\ G} + \frac{1}{K} \Rightarrow \tag{4.22}$$

$$E_{cv} = \frac{K\ (6\ G\ S_1^2 + 3\ G)}{6\ G\ S_1^2 + 3\ G + K}$$

The above expressions may be used in the design of elastomeric bearings, according to the significance of each component of deformation.

Another issue which must be addressed is the prevention of buckling, which may occur due to the reduction of the critical buckling load caused by the shear flexibility of the rubber. The buckling load of a slender column with length l and bending rigidity EI, either fixed supported or hinged at both ends, is given by the following two expressions:

$$P_{cr} = P_E = \frac{4\ \pi^2\ E\ I}{l^2}\ , \qquad P_{cr} = P_E = \frac{\pi^2\ E\ I}{l^2}$$

The reduction of the buckling load due to shear flexibility, for $P_{cr} = P_E$, is given by the following expression:

$$P_{cr} = \frac{\sqrt{1 + 4\ P_E/P_G} - 1}{2/P_G} \tag{4.23}$$

(Stanton and Roeder [99]), where $P_G = G\ A_s$ and A_s is the effective shear area.

Elastomeric bearings are very flexible in shear while very stiff in compression and thus the ratio P_E/P_G is large. As this ratio approaches infinity, the critical-for-buckling loading tends to

$$P_{cr} = \sqrt{P_E\ P_G} \tag{4.24}$$

A correction coefficient may be applied to consider the non-absolutely rigid, in flexure, steel laminates. The proposed (Stanton and Roeder [104]) correction coefficient f_r is equal to $f_r = B_r\ S_1^2$ where B_r is equal to 0.742 for a square bearing and 0.50 for a circular one. The critical load, for a circular bearing, therefore, with design details which avoid transmission of bending moments, is determined by

$$P_{cr} = \sqrt{\frac{\pi^2\ E\ \pi\ D^4}{h_r^2}\left(\frac{0.5\ D}{4\ t_r}\right)^2\frac{G\ \pi\ D^2}{4}} \tag{4.25}$$

When Equation 4.17 is substituted above, we obtain the following expression for the buckling load which may be used for a stability check.

$$P_{cr} = \sqrt{\frac{\pi^2\ 6\ G\ D^2\ \pi\ D^4}{h_r^2\ (4\ t_r)^2\ 64}\ 0.5\left(\frac{D}{4\ t_r}\right)^2\frac{G\ \pi\ D^2}{4}} \tag{4.26}$$

$$\Rightarrow P_{cr} \approx \frac{0.134\ G\ D^5}{h_r\ t_r^2}$$

Finally, another issue of concern is the vertical load capacity of a bearing which depends on the plan area of the bearing and the internal rubber sheet thickness. An effective column may conservatively be considered to transfer axial loads, neglecting the remaining area (Figure 4.10). The effective area A_{eff} may be computed by simple geometric considerations, as a function of the plan area A when the maximum displacement U_{max} is defined.

$$\theta = 2\ acos\left(\frac{U_{max}}{D}\right)\ , \qquad A_{eff} = \left(\frac{\theta}{180} - \frac{sin\theta}{\pi}\right) A \tag{4.27}$$

As shown in Figure 4.10, the U_{max}/D ratio must not be larger than 0.40 to have an effective overlapping area of more than half of the total plan area A.

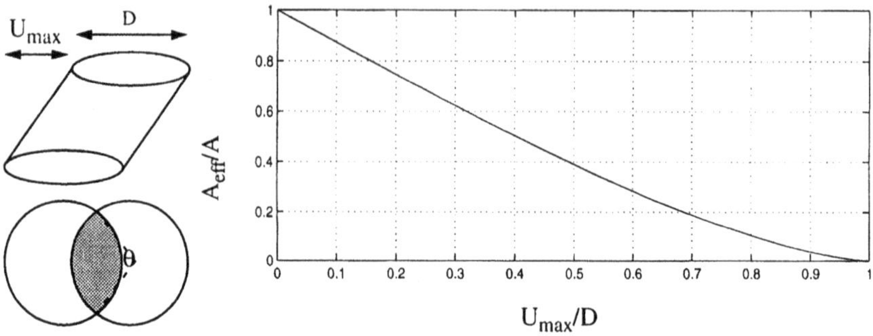

Figure 4.10: Effective column area reduction with displacement.

The shear stiffness of an elastomeric bearing without a lead plug can be estimated using the formula

$$K^b = \frac{G_r A_r}{h_r} \tag{4.28}$$

where G_r is the shear modulus of rubber, A_r is the cross-section area of the rubber, and h_r the total height of the rubber.

The shear stiffness of an LRB, prior to and after the yielding of the lead, can be approximated by the following two formulas:

$$K^b_{el} = \frac{G_l A_l}{h_l} + \frac{G_r A_r}{h_r} \quad , \qquad K^b_{pl} = \frac{G_r A_r}{h_r} \tag{4.29}$$

where G_l is the shear modulus of lead, A_l is the cross-section area of the lead plug(s), and h_l is the height of the lead. In Equation (4.29), A_r is the net cross-section area of the rubber, i.e. the gross area of the bearing minus the cross-section area of the lead plug(s). The presence of the lead plug for the post-yield stiffness of the LRB has been ignored.

The above formulas can be used as approximations for the preliminary design of a seismically isolated structure. However, they are based on pure shear deformations of the rubber and neglect the effect of axial forces and deformations. In the final design and analysis process, the influence of the normal stresses and strains may need to be taken into account, as well as the presence of the lead after its yielding and its contribution to the shear stiffness. The dependence of the horizontal stiffness on axial load can be approximated (Buckle and Kelly [17]) by

$$K_h^* = K_h\left[1 + \left(\frac{P}{P_{cr}}\right)^2\right] \tag{4.30}$$

where P is the axial load, P_{cr} is the critical load, K_h is the horizontal stiffness under zero axial load, and K_h^* is the effective horizontal stiffness under axial load P. However, the critical load P_{cr} may decrease with increasing horizontal displacement. This reduction of the critical load, as well as the dependence of the horizontal stiffness on both horizontal displacement and vertical load, may need to be taken into account in a more detailed non-linear model at the final analysis stage. The dependence of the shear modulus of rubber $G_r(\gamma)$ on its shear strain, γ, can also be taken into account by means of a non-linear and more detailed model.

According to experimental work on the behavior of elastomeric bearings used for seismic isolation under shear and axial loads (Mori et al. [85]), using the linear elastic theory to calculate the shear stiffness does not give an accurate prediction of performance. It was also found (Mori et al. [85]) that under very large shear strains (of the order of 200%), lift-off action at the bearing edges had occurred and the maximum normal stresses were significantly higher than the average compressive stresses on the bearings. Although the experiments indicated that seismic isolation bearings can withstand very large shear deformations for short-term loading levels, the steel plates bend at such shear strain levels. This indicates that if elastomeric bearings are subjected to very large shear strains during an earthquake they may need to be replaced afterwards.

In addition, even when design details do not allow the transfer of bending moments to the bearings, the distribution of compressive stresses is not uniform, due to the shear force and the resulting overturning moment which must be counterbalanced. Thus, either the maximum compressive stresses, i.e. those at the edges, must be checked, or appropriate safety factors must be used for average compressive stresses. Experiments on the behavior of elastomeric bearings used for seismic isolation under rotation and axial loads (Mori et al. [86]) showed that the rotational stiffness of an elastomeric bearing is independent of the axial load as long as the latter is not so large as to cause nonlinear behavior of the bearing in compression.

Finally, elastomeric bearings are not able to withstand large tensile stresses, which may be caused by overturning moments in high-rise buildings. Therefore, special considerations must be taken into account when seismically isolating a high-rise or aspect ratio building. Most seismically isolated buildings have a height of less than 50 meters and an aspect ratio of less than 3. When pull-out forces on the bearings are expected, either design details can be used to prevent tensile stresses and strains in the elastomer, or special high-strength elastomer bearings can be used. High-strength rubber bearings can be manufactured by mixing carbon with rubber, which increases its compressive and tensile strength. Fiber-reinforced elastomeric bearings can provide high-tensile strength and withstand tensile stresses.

4.4.3 Available evaluation tests of seismic isolation and energy dissipating devices

In 1994, the Highway Innovative Technology Center (HITEC), the Federal Highway Administration (FHWA), and the California Department of Transportations (Caltrans) cooperated to perform full-scale dynamic tests and evaluations of seismic isolation and energy dissipation devices. The evaluation program aimed at verifying the mechanical and performance properties of the systems under evaluation. The guidelines for the testing were set out in a report published by the Highway Innovative Technology Center (HITEC [42]). The cooperative program also aimed at the provision guidelines on the selection and design of these systems, as well as the development of specifications for their use in new or existing bridge structures.

The performed tests verified the stiffness and dissipating properties of the systems, evaluated the effects of compressive loads on these properties, estimated their dynamic properties under varying frequencies, evaluated their fatigue, wear, and environmental aging properties, and assessed the effects of extreme temperature on their performance characteristics and their performance under ultimate displacements (HITEC [42]).

Although the main objective was to provide guidance for the use of seismic isolation for bridges, the results and conclusions of the tests are valuable for the application of seismic isolation to any kind of structure. The advantages of each system can be considered using the results of these impartial tests. In addition, the full-scale dynamic tests help in the establishment of greater reliability for seismic isolation devices.

Eleven systems were tested, among them lead-rubber bearings, high-damping bearings, steel rubber bearings, slider bearings, roller bearings, friction pendulum bearings and viscous dampers. The results for each evaluated system are presented in individual reports (HITEC [43]-[53]) released by the Civil Engineering Research Foundation (CERF). CERF is an independent non-profit organization, a subsidiary of the American Society of Civil Engineers (ASCE).

Chapter 5

Applications of Seismic Isolation

5.1 Introduction

The concept of decoupling a structure from the ground, via a mechanism on which the structure is supported, dates back to the early 1900s (Buckle and Mayes [18]; Naeim and Kelly [88]). Initially, layers of sand or a similar material at the foundation level had been proposed, to allow slippage of the structure during a severe earthquake. The concept of seismic isolation resulted from observations of structures which were accidentally built on soft soil layers and which experienced limited damage during severe earthquakes. These structures slipped as rigid bodies on the soft soil layers. In contrast, adjacent fixed supported structures experienced severe damage or even collapsed. Since then, several seismically isolated structures have been built, either intentionally or accidentally. The fact that in most cases these structures survived severe earthquakes, while adjacent structures collapsed, established seismic isolation as a way of resisting earthquakes.

The first known proposal to isolate a structure from the movements of the earth was reported in Japan in 1891. The first official reference to seismic isolation was made in 1906, in the United States, by a patent application for an earthquake-proof building mounted on rollers. In the same period, the idea of resisting the actions of ground disturbances on buildings by founding them on layers of talc for isolation purposes was proposed in England (Kelly [56]; Naeim and Kelly [88]). In 1908 in Italy, a proposal was also made for the reconstruction of an earthquake damaged area, using rollers or layers of sand to separate each building from its foundation (Augenti and Serino [11]). Many methods have been invented and used in order to uncouple a structure from damaging earthquake motion, essentially seismically isolating the structure. During the last quarter of this century there has been an increasingly wide use of seismic isolation, which has coincided with the development and broad availability of elastomeric bearings. The awareness and increased sensitivity of the public to the tragic consequences of recent powerful earthquakes has helped the establishment and wider use of seismic isolation as a way of mitigating seismic hazards. Human casualties and extensive damage due to a strong earthquake are no longer acceptable since they can now be avoided or at least limited. Today, more than four hundred isolated structures have been built and the rate keeps increasing. However, most applications of seismic isolation, considering new constructions, are important structures which house sensitive, expensive or critical equipment. Seismic isolation has rarely been used for common public housing construction projects, although some proposals for low-cost seismic isolation, specifically for such cases, have been made.

In addition, a potential area for the application of seismic isolation is the seismic upgrading of existing structures, especially bridges. The rehabilitation and seismic performance upgrading of historical structures of architectural importance using seismic isolation may be accomplished with many fewer modifications than would be necessary using any conventional strengthening method. The effectiveness of seismic isolation in retrofitting existing structures is based on reducing the seismic forces on the superstructure instead of strengthening the structure to withstand high seismic forces. Some stiffening of the superstructure may also increase the effectiveness of seismic isolation.

The method is also attractive for seismic upgrading of existing bridges since the isolators can easily be installed by simply replacing standard bearings, used for thermal expansion, with seismic isolation ones. The isolation bearings are usually inserted between the tops of the piers and the superstructure of the bridge. In addition, it is necessary to provide a gap to accommodate the larger relative displacements. The most effective way to isolate an existing bridge is to use lead rubber bearings (LRBs), since they combine the required initial rigidity for minor loads, e.g. the horizontal forces from car braking, flexibility to isolate the superstructure, restoring force to bring the structure to its original configuration and a dissipation mechanism, in one compact unit, while supporting the weight of the structure.

5.2 Seismic Isolation in the United States

In the United States the number of seismically isolated bridges is about four times the number of seismically isolated buildings. The latter are almost equally divided between new constructions and retrofits of existing buildings. Most seismically isolated structures in the United States house critical, or expensive and valuable, contents, such as hospitals and high-technology centers.

The number of seismically isolated structures is steadily increasing. for example, in 1989 only five buildings and six bridges had been seismically isolated, while by 1993 another twenty buildings and fifty bridges (Figure 5.1) had been seismically isolated (Mayes [72]).

Figure 5.1: Increase of seismically isolated buildings and bridges in the USA.

Today in the United States, the number of seismically isolated buildings is estimated to be over 50 and the number of seismically isolated bridges is estimated to be over 200.

5.2.1 The Foothill Communities Law and Justice Center

The Foothill Communities Law and Justice Center (Figure 5.2) in San Bernardino, a four-story concentrically braced steel-frame building, was the first seismically isolated structure built in the United States. The building, which cost $30 million, was mounted on 98 high-damping rubber isolators, becoming the first seismically isolated building in the world where high-damping elastomeric bearings (HDRBs) were used.

Figure 5.2: Foothill Communities Law and Justice Center, California, USA.

The building, approximately 15,800 m^2, is located in a high seismicity area, 21 km from the San Andreas fault and about 100 km from downtown Los Angeles. The superstructure consists of steel frames stiffened by brace frames in some bays of the building. It has four stories, a basement, and a sub-basement where the isolation system is located. An almost 40 cm gap was provided for the isolators at the corners of the building to accommodate a large displacement at the isolation level taking into account potential torsional effects.

The construction began in 1994 and took a little over a year to complete. The building was designed and built to withstand an earthquake of 8.3 magnitude on the Richter scale. During the 1990 Upland earthquake, which had a 5.5 Richter magnitude, there was no damage to the building.

5.2.2 The Salt Lake City and County Building

The Salt Lake City and County Building is the first historic structure to have been seismically isolated. The five-story, 80x40 m in plan view building, which weighs approximately 34,000 tons, was built at the end of the nineteenth century out of unreinforced brick, masonry and sandstone (Figure 5.3).

Figure 5.3: Salt Lake City and County Building, Utah, USA.

A twelve-story, 69 m high and 4 m^2 plan view area clock tower, located at the center of the building, was also built out of unreinforced masonry. Seismic isolation was provided, using 447 elastomeric bearings installed between the structure and its massive foundation system (Bailey and Allen [12]). Lead rubber bearings were used under the exterior walls to provide high initial rigidity in order to account for wind loads. A reinforced retaining wall around the structure provided a 30 cm gap to accommodate the relative displacement at the level of the isolation system.

Seismic isolation was selected as the least disruptive of different methods, in terms of architectural aesthetics, and the most effective to minimize damage from a strong seismic excitation among three alternative retrofit schemes. The disruption of activity in the masonry superstructure was minimized due to the substantial reduction of the induced seismic forces and, consequently, the limited stiffening that was required. Seismic isolation was incorporated at a comparable cost to a conventional, according to the Uniform Building Code (UBC), strengthening scheme. The structure is located in a high seismicity area, situated within two miles from the Wasatch fault. Seismic isolation reduces the induced seismic forces by a factor of six, thus eliminating the need for conventional strengthening of the building (DIS [31]). Computer simulations indicated that the structure would remain elastic under the design seismic excitation of 0.20 g peak ground acceleration and would not, therefore, be damaged. In contrast, the analysis indicated that it would be impossible for the superstructure to remain elastic if conventional strengthening methods were to be used instead of seismic isolation.

5.2.3 The University Hospital of the University of Southern California

The USC University Hospital, shown in Figure 5.4, is an eight-story, 32,500 m^2 building at the University of California Medical Campus. It is the world's first seismically isolated hospital. It is also the first seismically isolated building in the United States that has experienced a strong earthquake excitation, namely the Northridge earthquake in 1994. Its performance during that earthquake is described in detail in Paragraph 5.6.1 on page 79.

Structural costs for both conventional and isolated options were considered and compared in economic terms. The comparison indicated that the savings on the structural frame, if seismic isolation were incorporated, counterbalanced the additional cost of the isolation system and the additional slab at the isolation level, placed to ensure uniform distribution of the lateral forces at the isolators isolation was provided, using 447 elastomeric bearings installed between the structure and its massive foundation system (Bailey and Allen [12]). Lead rubber bearings were used under the exterior walls to provide high initial rigidity in order to account for wind loads. A reinforced retaining wall around the structure provided a 30 cm gap to accommodate the relative displacement at the level of the isolation system.

Seismic isolation was selected as the least disruptive of different methods, in terms of architectural aesthetics, and the most effective to minimize damage from a strong seismic excitation among three alternative retrofit schemes. The disruption of activity in the masonry superstructure was minimized due to the substantial reduction of the induced seismic forces and, consequently, the limited stiffening that was required. Seismic isolation was incorporated at a comparable cost to a conventional, according to the Uniform Building Code (UBC), strengthening scheme. The structure is located in a high seismicity area, situated within two miles from the Wasatch fault. Seismic isolation reduces the induced seismic forces by a factor of six, thus eliminating the need for conventional strengthening of the building (DIS [31]). Computer simulations indicated that the structure would remain elastic under the design seismic excitation of 0.20 g peak ground acceleration and would not, therefore, be damaged. In contrast, the analysis indicated that it would be impossible for the superstructure to remain elastic if conventional strengthening methods were to be used instead of seismic isolation.

5.2.3 The University Hospital of the University of Southern California

The USC University Hospital, shown in Figure 5.4, is an eight-story, 32,500 m^2 building at the University of California Medical Campus. It is the world's first seismically isolated hospital. It is also the first seismically isolated building in the United States that has experienced a strong earthquake excitation, namely the Northridge earthquake in 1994. Its performance during that earthquake is described in detail in Paragraph 5.6.1 on page 79.

Structural costs for both conventional and isolated options were considered and compared in economic terms. The comparison indicated that the savings on the structural frame, if seismic isolation were incorporated, counterbalanced the additional cost of the isolation system and the additional slab at the isolation level, placed to ensure uniform distribution of the lateral forces at the isolators lead plug cores. The dimensions of the bearings varied from 740 to 940 mm in diameter and 400 mm in height. The isolation system was supported on a thick concrete mat foundation. A continuous 500 mm gap had also been provided, by constructing a moat wall around the building, to ensure the accommodation of large relative displacements at the isolation level.

The cost of the isolators was around $2.1 million, while the total cost of the retrofitting was around $84 millions. A 0.48 g S_2 was considered as the design basis carthquakc The scismic rchabilitation of thc structure was completed in 1995, the same year the seismic retrofitting of the San Francisco City hall had started.

5.2.5 San Francisco City Hall

With 500 elastomeric isolators installed, the San Francisco City Hall is the largest seismic retrofitting project in the world. The structure, which was built in 1912, had been severely damaged by the 1989 Loma Prieta earthquake. The objective was to preserve the architectural characteristics of the building while increasing its seismic capacity. It was decide that seismic isolation be used to seismically upgrade the five-story (90 m tall) steel-frame structure.

Figure 5.6: San Francisco City Hall, California, USA.

(Mayes et al. [71]). The provided seismic gap around the building is 26 centimeters.

Figure 5.4: University Hospital of the University of Southern California, USA.

The steel perimeter braced-frame structure was completed in 1991. It was mounted on 68 lead rubber and 81 elastomeric bearings. At the exterior columns of this asymmetric-plan building lead rubber bearings were used, while in the interior columns elastomeric bearings without lead plugs were used. The structure was designed to withstand an 8.2 earthquake on the Richter scale.

5.2.4 Seismic retrofit of Oakland City Hall using seismic isolation

The 18-story, 99 m high and approximately 15,000 m^2 Oakland City Hall, which was constructed in 1914, was the first high-rise building in the United States. It was also the tallest seismically isolated building in the world (Elsesser et al. [35]) at the time of the completion of its seismic rehabilitation. Interest in seismic retrofit and upgrading of the steel-frame building with infill walls was expressed, after the extensive damage it experienced during the Loma Prieta earthquake. In the vertical direction, the building is composed of a three-story base, a ten-story office tower, a two-story clock tower base and a three-story clock tower, as shown in Figure 5.5.

The building was seismically isolated using 530 lead-rubber bearings. Concrete shear walls were also used to strengthen and stiffen the superstructure. New retaining walls had to be built around the building to provide the required seismic gap between the structure and the ground.

5.2.6 Los Angeles City Hall

The Los Angeles City Hall (Figure 5.7) is an historic 32-story and 83,000 m^2 building that was constructed in 1926 without any earthquake resistance requirements. The structure experienced substantial damage from various strong earthquakes to the extent that its preservation was questioned. Severe damage occurred during the Northridge earthquake in 1994. The city decided to save the building by seismically upgrading it. The objective was to enhance the earthquake resistance of the structure while preserving its unique architectural characteristics and interior design details.

Figure 5.7: Los Angeles City Hall, California, USA.

Three different schemes for seismically upgrading the building were considered. Two of the schemes were conventional strengthening schemes and the third was seismic isolation. The conventional strengthening schemes required the addition of structural elements to increase the strength and stiffness of the building, which, however, would have resulted in greater seismic demands. In contrast, the seismic isolation scheme would reduce the seismic forces transferred to the structure by decoupling it from the moving ground during a strong earthquake excitation.

Overall, seismic isolation was found to be more effective in terms of performance and cost than the two conventional strengthening schemes. The evaluation of the three different schemes focused on the inter-story drift and the floor accelerations. While all three schemes reduce the inter-story drift, only seismic isolation reduces the floors accelerations to acceptable levels. High floor accelerations could result in damage of the nonstructural components of the building and its contents, as well as potential hazards due to falling debris.

The construction started in 1997 and will have lasted for four years. The structure will have been seismically isolated using 416 high damping rubber bearings (HDRBs) and 90 sliding bearings. Upon completion of the structure, it will be the tallest seismically isolated structure in the world.

The isolation plane will be located between the existing foundation and the basement level. A loose connection detail will be used at each isolator to allow its limited uplift without loading the bearing in tension to avoid problems due to its limited tension capacity. A twenty-centimeter reinforced concrete diaphragm will replace the existing basement diaphragm in order to tie together all isolators and ensure proper load transfer distribution between the isolation system and the superstructure. There will be some strengthening of the existing structure and its foundations to provide the necessary strength and stiffness as well as to improve the load transfer between structural elements.

In addition, to increase the energy dissipation capacity of the structure, 52 viscous dampers will be added at the isolation level. Another twelve dampers will be added in the tower between the 24th and the 25th floors to provide damping for the higher modes.

5.3 Seismic Isolation in Japan

In Japan, the first seismically isolated structure was a two-story reinforced concrete dwelling. It was mounted on six elastomeric bearings inserted at the foundation level under its six columns. The first large seismically isolated building, a four-story reinforced concrete structure, was completed in 1986. Today, there are more than 75 seismically isolated buildings and 20 bridges, the majority in Japan located in the Tokyo area. The last few years have seen a significant increase in the use of seismic isolation. The number of applications to the Ministry of Construction of Japan for use of seismic isolation is of the order of several hundreds. The number of seismically isolated buildings that had been approved by the Ministry of Construction up to mid-1998 exceeded 500 (Kelly [61]). To construct a seismically isolated structure in Japan requires a review and approval from a standing committee of the Ministry of Construction. It is estimated that the number of permits granted for the use of seismic isolation will be around 200-300 per year, especially after observation of the superior performance of seismically isolated structures during recent powerful earthquakes.

The majority of seismically isolated structures in Japan are residential and office buildings. It is evident that the Japanese are more willing to pay the extra initial cost for seismic isolation in order to achieve higher levels of seismic protection. Seismic isolation is relatively more rarely used for seismic retrofitting of existing structures, in contrast to US practise where seismic isolation is very commonly used in retrofitting of buildings of historical value.

The seismic isolation concept had been proposed initially in 1924, in Japan, with a ball-bearing system (Kitagawa [63]). In the late 1970s, partial isolation was used in more than one hundred continuous bridges, in combination with viscous dampers. The most commonly used system in fully seismically isolated structures is one consisting of lead rubber bearings or natural rubber bearings with auxiliary mechanical dampers. Today, there is an increasing trend towards the use of high-damping rubber bearings.

Many isolated structures are fully instrumented to record earthquake excitations. Thus, due to the frequent occurrence of seismic events in Japan, important information has been collected for the performance of seismically isolated structures. Whenever the response of an isolated structure has been recorded and compared with that of an adjacent fixed-base structure, the comparison has confirmed the superiority of seismic isolation, especially for ground motions with high acceleration levels (Kelly [58][59]). This has led to a wider use of seismic isolation in Japan. The West Japan Postal Computer Center, which is currently the world's largest seismically isolated building with an area of 47,000 m^2, experienced the 1995 Kobe earthquake. A detailed description of the building, and data regarding its response under that earthquake, are provided in Section 5.6.2: The West Japan Compute Center.

5.3.1 The Computer center of the Tohoku Electric Power Company

The computer center of the Tohoku Electric Power Company is located in Sendai, Miyako Province. The six-story computer center was constructed to house expensive and critical equipment belonging to the electric power utility (Figure 5.8). Upon its completion in 1990, it was the largest seismically isolated building in Japan with 10,000 m^2 floor space.

Figure 5.8: The Tohoku Electric Power Computer Center, Japan.

The building was mounted on 40 high-damping bearings with diameters from 0.90 to 1.20 m, each carrying vertical loads of between 400 and 800 tons. The seismic isolation system cost an additional 5% on top of $20 million structural cost. Its installation was very simple and the construction of the building took only one year.

5.3.2 The C-1 Building

This building (Figure 5.9) houses a computer center and was completed in 1992 in Fuchi City, Tokyo. It was the largest seismically isolated building in the world when it was completed in 1992.

Figure 5.9: The C-1 Building, Tokyo, Japan.

The building is a seven-story, composite reinforced concrete-steel structure of 41.4 m maximum height above ground level, 62,800 tons total weight and 37,850 m^2 total isolated superstructure floor area. Expensive and sensitive equipment is stored in the building. The protection of the building's contents was the dominant design requirement. The superstructure is mounted on 68 circular lead rubber bearings of 1.1 to 1.5 m diameter, with lead core plugs of 180 to 200 mm diameter. The 1.5 m diameter bearings are the biggest lead rubber bearings used in the world (Nakagawa and Kawamura [91]). The isolators carry large loads since they are installed on a 15 m by 15 m grid. The fundamental period of the isolated structure varies from 1.4 to 3.0 seconds depending on the intensity of the excitation. A dynamic analysis was performed for three levels of maximum ground velocity: the standard 25 cm/sec, 50 cm/sec, and 75 cm/sec for fail-safe characteristics check. A fundamental period of 3.0 seconds corresponds to the second level of input. The base shear due to design wind loads is 43% of the yield force of the isolation system, ensuring its safety against wind loads.

5.4 Seismic Isolation in New Zealand

Together with Japan, New Zealand has pioneered the concept of seismic isolation. Although a type of seismic isolation was proposed as early as 1929 (Buckle and Mayes [18]), practical implementation of the method began in 1973 with the construction of an isolated bridge (McKay et al. [77]). The first seismically isolated building, the William Clayton Building in Wellington, was constructed in

1981. The lead rubber bearing system is the most commonly used seismic isolation system in New Zealand, especially in bridge construction where its use clearly dominates over that of other systems.

Today, it is estimated that there are about ten seismically isolated buildings and more than fifty seismically isolated bridges in New Zealand. The increasing number of seismically isolated structures in New Zealand indicates the wide acceptance of this innovative concept for protection against earthquakes in an area of very high seismicity.

5.4.1 The William Clayton Building

The William Clayton Building in Wellington (Figure 5.10) was the first seismically isolated building in New Zealand and the first in the world to incorporate lead rubber bearings. Its construction started in 1978 and was completed in four years (McKay et al. [77]).

Figure 5.10: William Clayton Building, New Zealand.

The four-story reinforced concrete frame building, (97 m by 40 m in plan view), is mounted on 80 square lead rubber bearings. Each column of the structure is supported on a 600 mm by 600 mm by 207 mm tall bearing with a 105 mm diameter lead cylindrical plug. The vertical load carried by each bearing varies from 1.0 to 2.0 MN. A gap around the structure and special details of non-structural elements and internal services crossing the isolation level allow 150 mm relative displacement in any horizontal direction. The effective fundamental period of the building is between 0.8 and 2.0 seconds depending on the level of applied forces and corresponding displacements, while the level of the yield forces is 7%

of the weight of the building. The design of this isolated structure was conservative due to the proximity of the building to the active Wellington fault and the limited experience of seismic isolation at the time of its construction.

5.4.2 Wellington Central Police Station

This ten-story, 38 m by 31 m reinforced-concrete building was seismically isolated to ensure that it will remain fully operational in a major earthquake and fulfill its special functions in such emergencies (Figure 5.11). Three different construction schemes were initially considered (McKay et al. [77]); seismic isolation finally dominated with 10% savings on the structural cost and with a higher earthquake resistance capacity. The conventional earthquake resistance schemes would have required large structural elements to meet the rigorous strength and deflection requirements for such a special building.

Figure 5.11: Wellington Central Police Station, New Zealand.

Seismic isolation could not be implemented with elastomeric bearings due to the height and the plan dimensions of the building. Thus, a different isolation scheme was considered. Long piles enclosed in large-dimension casings, to provide an horizontal separation gap from the surrounding soil, were used combined with lead extrusion dissipators. The latter are hysteretic energy dissipation mechanisms based on the plastic deformations of lead during its change in shape when forced or extruded through a hole. The piles were also necessary for the foundation of the building to reach the bedrock below the 15 m layer of weak soil deposits. The structure is located within a few hundred meters of the active Wellington fault. The adopted seismic isolation scheme not only resulted in considerable savings on the initial cost, but also ensured that the building would remain operational during a severe earthquake.

5.4.3 New Zealand Parliament Buildings

The $165 million project to strengthen New Zealand's parliament is the largest such project in New Zealand (Gifford [41]). Two historic New Zealand buildings, the Parliament House built in 1912 and the Parliamentary Library built in 1898, have to be seismically strengthened to be able to withstand major earthquakes and to avoid damage and human casualties. Strengthening is imperative, considering the high seismicity of the site and the proximity of the buildings (within 400 meters) to the nearby Wellington active fault as well as to other active faults. The historical and architectural value of the two structures (Figure 5.12) gives great importance to the preservation of their architectural character.

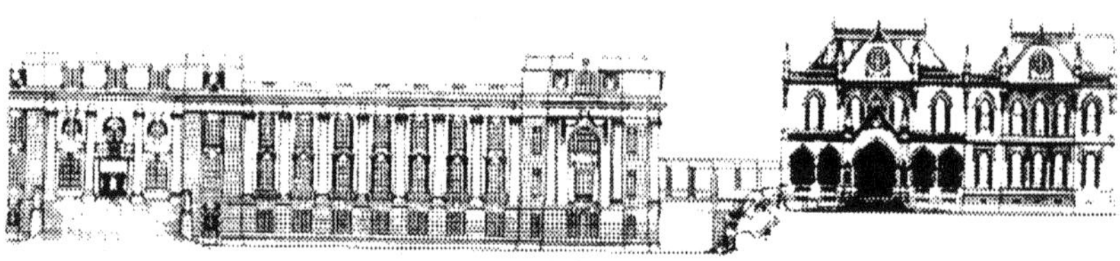

Figure 5.12: Parliament Buildings, New Zealand.

Seismic isolation has been selected among several alternative strengthening design schemes, due its limited impact on architectural characteristics. Both buildings will be placed on rubber bearings, mainly high-damping rubber bearings, most of them having lead cores (LHDRB). In particular, more than 75% of the four hundred bearings will be lead high-damping rubber bearings (LHDRBs). It will be the first time that this hybrid isolation system, which has been described in detail in Chapter 4, will have been used.

The isolation system will be placed at the lower level of the five-story Parliament House, while floor diaphragms and internal walls will be stiffened and columns and walls in the basement will be strengthened. The installation of the isolators was enabled by gradually cutting large seven-ton blocks out of the existing foundations, and building new reinforced-concrete foundations (Figure 5.13).

Figure 5.13: Cutting out existing foundations to build new foundations on which seismic isolators can be installed, New Zealand's Parliament Buildings.

The Parliamentary Library will be seismically isolated by an hybrid isolation system installed at its basement. Concrete facing will be added to the non-architectural side of selected walls. Over four hundred bearings will be inserted in the foundation walls at basement level to provide the required flexibility. Numerical simulations of the hybrid isolation system and the isolated structure indicated the superiority of the hybrid system over common lead rubber bearings. In addition, experimental tests have been performed with successful results, to ensure matching of the stiffness characteristics of the analysis model and the original bearings. The fundamental periods of both buildings will be lengthened from approximately 0.50 to 3.5 seconds, reducing substantially the seismic forces and inter-story deflections (DIS [31], [32]). Although the alternative conventional strengthening resistant option was 3% cheaper, seismic isolation was preferred, since it provided a higher earthquake resistance capacity and limited impact on the valuable architectural characteristics of both buildings. It is estimated that seismic isolation will enable these masonry-bearing wall structures to withstand earthquakes of up to 7.5 magnitude on the Richter scale.

5.5 Seismic Isolation in Italy

In Europe, seismic isolation has been used mostly in Italy, especially in bridges of the Italian highways. In Italy, the GLIS (Gruppo de Lavoro Isolamento Sismico), which is the National Working Group on Seismic Isolation, was established in 1989 to support the research and implementation of seismic isolation. GLIS performs numerical and experimental studies on seismic isolation, provides technical support to designers of seismically isolated structures, promotes the use of seismic isolation, supports the development of relevant codes and standards, organizes conferences on seismic isolation, and supports State Institutes to verify and approve proposed seismic isolation projects. The role of GLIS is very important in the promotion and wider use of seismic isolation.

It is estimated that there are more than 30 seismically isolated buildings and about 200 seismically isolated bridges in Italy. The most commonly used seismic isolation system is high-damping rubber bearings (HDRBs).

One seismically isolated building is the SIP Complex, which is the new administration center of the National Telephone Company in Ancona (Figure 5.14).

Figure 5.14: The SIP Complex, Ancona, Italy.

This seven-story building is mounted on 297 high-damping rubber bearings (HDRBs) of diameters 0.5 to 0.6 m and height 0.19 m. The construction of the SIP Complex started in 1989 and was completed in 1992.

5.6 Observed Performance of Seismically Isolated Structures

Several isolated structures are instrumented with equipment to record their response and thus provide information on their performance. As the number of isolated structures increases, more information is collected from isolated structures subjected to strong ground motions. The recorded information consistently indicates reduced accelerations above the isolation level in comparison to adjacent conventionally designed or strengthened structures. The latest available information related to seismic isolation is from two recent major earthquakes, the 1994 Northridge earthquake in California and the 1995 Kobe earthquake in Japan.

In particular, the following sections present information that has been reported on the performance of the University of Southern California Hospital, subjected to the Northridge earthquake (1994); the West Japan Computer Center of the Ministry of Post and Telecommunications, subjected to the Kobe earthquake (1995); the Eel River Bridge in California, subjected to the 7.0 Richter magnitude Petrolia earthquakes (1992); and the Rangitaiki River Bridge in New Zealand, subjected to the 6.3 Richter magnitude Edgecumbe earthquake (1987).

5.6.1 The University Hospital of the University of Southern California

The January 17, 1994, Northridge earthquake of 6.7 magnitude on the Richter scale resulted in 61 deaths and more than 9000 injured people. The earthquake casualties could have been much higher if the earthquake had happened during normal business hours instead of occurring at 4:31 a.m. The damage cost is estimated to be between 20-50 billion dollars, while the replacement value of damaged property is estimated to be about 800 billion dollars. The earthquake significantly damaged numerous buildings in the Los Angeles area, including 31 hospitals (DIS [31], [32]). The USC University Hospital, located 36 km from the earthquake epicenter, remained operational with absolutely no damage, while many conventionally designed earthquake-resistant hospitals suffered extensive damage. This was due to the fact that the earthquake ground accelerations, and consequently the resulting seismic forces, were reduced by 65%. More information regarding the USC University Hospital has been provided in Paragraph 5.2.3 on page 65.

Preliminary accelograms, which have been released by the California Strong Motion Instrumentation Program (SMIP), demonstrate the effectiveness of seismic isolation and encourage its use. The variations of ground and floor accelerations are shown in Figure 5.15 (DIS [33]).

Figure 5.15 shows the peak recorded accelerations of the ground and certain floors of the hospital. It is evident that instead of amplifying the peak ground accelerations, the structure had smaller floor accelerations than the recorded peak ground acceleration (pga). The peak free-field and peak foundation accelerations in the north-south direction, in which the strongest motions were recorded, were 0.49 and 0.37 g, respectively. The maximum floor accelerations were recorded at the base and the roof of the building to be 0.13 and 0.21 g, respectively (Moehle [82]). The corresponding attenuation factors of the ground motion were 0.35 and 0.57.

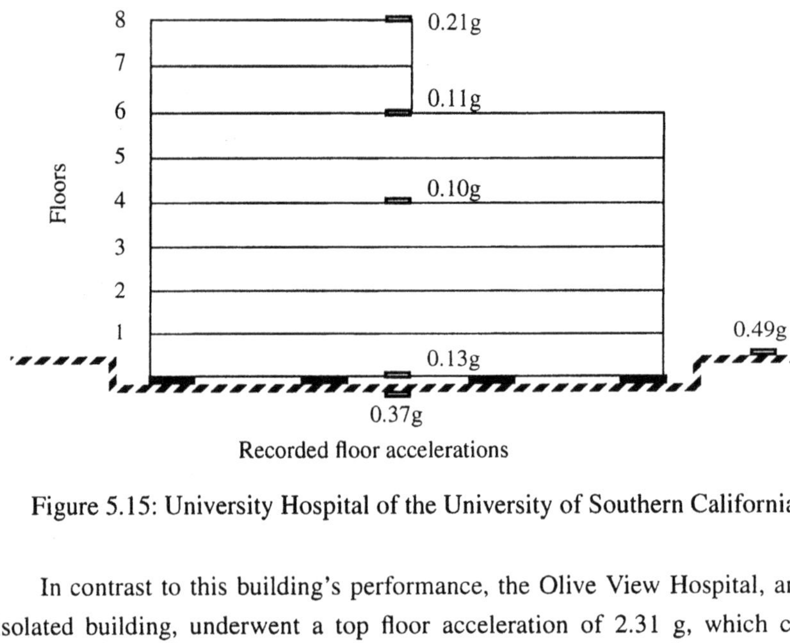

Figure 5.15: University Hospital of the University of Southern California.

In contrast to this building's performance, the Olive View Hospital, an un-isolated building, underwent a top floor acceleration of 2.31 g, which corresponded to a 2.8 amplification of its 0.82 g base acceleration. In essence, the ratio of the amplification factor of the Olive View Hospital to that of the University Hospital of the University of Southern California was equal to five ($2.8/0.57 \approx 5$).

Although free-field vertical ground accelerations as high as 0.85 g were recorded during the Northridge earthquake, recorded free-field vertical ground accelerations at the USC University Hospital did not exceed 0.09 g (Bozorgnia et al. [16]). The peak ground accelerations of the two vertical recordings at the foundation of the building, which are located below the plane of isolation, were

0.07 and 0.09 g. The corresponding maximum vertical accelerations recorded in the building were 0.08 g and 0.13 g, respectively, which corresponded to 114% and 144% amplifications of the corresponding peak vertical ground accelerations. The recordings and integrated displacements (Lew and Hutson [67]) indicated that the vertical motions were not greatly amplified by the isolation system, and the response of a seismically isolated structure is not significantly altered by vertical ground motions.

While the USC University Hospital was successfully protected by seismic isolation from damage or disruption of its operation, ten nearby hospitals had damage levels such that they had to be evacuated. The Los Angeles County General Hospital Complex, located within a kilometer of the USC University Hospital, suffered $400 million worth of damage; two wings, significantly damaged, had to be evacuated and remained permanently closed. Twelve hospitals suffered major damage and had to interrupt their normal operations and services. Some of these hospitals were built according to the latest earthquake resistance design standards and codes.

Finally, the observed performance of this seismically isolated structure demonstrated the anticipated restoring capabilities of the bearings that had been used. The structure returned to its original configuration after the end of the earthquake and the isolators re-centered at their original positions.

With an additional cost of about 2% of the overall construction work, the USC University Hospital was seismically isolated, hundreds of people were protected, very expensive equipment was saved, and the hospital remained operational during the critical hours following the powerful earthquake.

5.6.2 The West Japan Computer Center

One year after the Northridge earthquake, on January 17, 1995, an earthquake with a 7.2 Richter magnitude struck the city of Kobe in Japan. The earthquake caused a large number of casualties, exceeding 5,500 fatalities and more than 33,000 injured people. In addition, the earthquake caused severe damage, which was estimated to be worth more than $200 billion, i.e. 4% of Japan's GDP. Thousands of buildings of all types were heavily damaged, or collapsed. In particular, more than 150,000 buildings collapsed or were so heavily damaged that they needed to be demolished. The close proximity of the epicenter to an urban area (the densely populated Kobe), the shallow depth of the hypocenter and the severe ground motion contributed to the devastation caused by the earthquake. Today, Kobe is still recovering from the 1995 earthquake. Although many buildings have been demolished and reconstructed and many more have been repaired and strengthened, housing still remains a serious problem.

One of the world's largest seismically isolated structures, the West Japan Computer Center of the Ministry of Post and Telecommunications, was located 30 km from the epicenter of the 1995 Hyogoken Nanbu (Kobe) earthquake. The maintenance of the continued operations of the facility is very critical, since it serves as the computer center for all financial transactions in Western Japan. The isolated structure did not experience any damage (Moehle [83]), although it underwent severe ground motion. A nearby fixed supported building experienced some damage during the earthquake. The superior performance of this seismically isolated structure has motivated an increased use of seismic isolation, especially for apartment buildings.

Figure 5.16: West Japan Postal Computer Center, Japan.

The West Japan Computer Center (Figure 5.16) is a six-story, 46,500 m^2, seismically isolated reinforced-concrete building. The structure is supported on 120 elastomeric bearings, in particular 54 lead rubber bearings and 66 natural rubber bearings. In addition, steel and lead dampers were also used to increase its damping capacity. The isolated structure has a fundamental period of about 3 seconds, depending on the level of displacements at the isolation level. The structure was fully instrumented and the floor as well as the ground accelerations were recorded. The recorded maximum floor accelerations, which were much less than the ground accelerations, indicated the effectiveness of seismic isolation. The recorded maximum floor accelerations in both horizontal directions and the corresponding measured peak ground accelerations are shown in Figure 5.17.

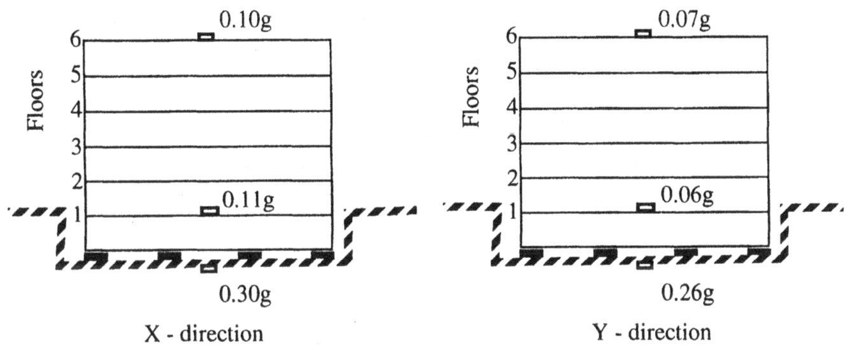

Figure 5.17: Recorded floor accelerations in the Computer Center, Kobe earthquake.

The ground accelerations were substantially reduced, whereas if the structure has been fixed supported, the floor accelerations would have been amplifications of the ground accelerations. The square root of the sum of the squares (SRSS) of the peak ground acceleration (pga) at the supporting ground, which equalled 0.41 g, was reduced to 0.31 g at the top of the building. This reduction in floor accelerations reduced not only the shear forces in the structural elements and the resultant damage, but also the damage to the contents of the structure due to accelerations.

A fixed supported building with similar structural characteristics would have had floor accelerations of an order of magnitude approximately higher. In particular, a nearby six-story fixed-supported reinforced-concrete structure experienced roof accelerations of 1.18 g (DIS [33]). It was demonstrated, therefore, that not only is the structure itself protected by seismic isolation, but also its contents, which in some cases, e.g. nuclear plants, is equally important to safeguard.

During the earthquake the building moved approximately 12 cm relative to the ground, and returned to its original configuration completely re-centering itself. The satisfactory re-centering demonstrated the restoring capabilities of elastomeric bearings.

The lessons learned from the Kobe earthquake and the observed superior performance of the seismically isolated structures, motivated an increased interest in seismic isolation. After the earthquake, many retrofitting schemes, especially for the seismic upgrading of highway bridges, have used seismic isolation, typically in combination with a strengthening of the superstructure.

An 18-span section of the Hanshin Expressway, which connects Osaka and Kobe, collapsed (Figure 5.18(a)) and had to be replaced. After the traffic had been rerouted along alternative detours, the collapsed viaduct was removed and new reinforced concrete piers combined with seismic isolators were constructed (Figure 5.18.b).

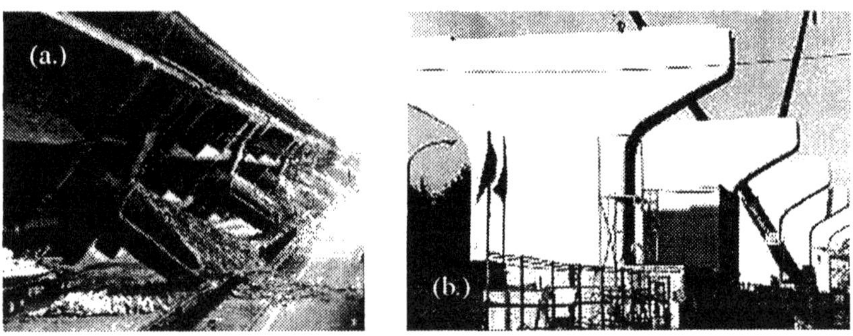

Figure 5.18: Viaduct section of the Hanshin Expressway.

At each pier, two lead rubber bearings of height 0.24 m and 1.1 by 1.1 m plan size were used. Each bearing had four lead cores, each with a diameter of 14 cm.

More than 60% of the bridge structures in Kobe were damaged by the 1995 earthquake. Since then, many regular steel bearings have been replaced with seismic isolation bearings during the reconstruction or retrofitting of the damaged bridge structures and viaducts. The latter have proven their effectiveness in reducing the seismic forces acting on the superstructure during a strong earthquake, while under normal conditions they can function as regular bearings. Typically, the seismic isolation bearings are placed between the tops of the piers and the superstructure.

5.6.3 The Eel River Bridge

The Eel River Bridge near Rio Dell, California, was built in 1940 and retrofitted in 1987. The bridge is located on Route 101 in Northern California and has two 92 meter spans. By seismically isolating the bridge, the base shear forces that can be transmitted to the bridge's superstructure have been reduced so much that there is no need for a costly upgrading and strengthening of the older piers. The seismic isolation was implemented by replacing the regular steel bearings with elastomeric bearings.

Five years later, on April 25, 1992, the 7.0 Richter magnitude Petrolia earthquake, with its epicenter located 14 miles from the Eel River bridge, gave the largest peak ground acceleration ever recorded in California, 2.0 g. Although the bridge was not instrumented, nearby instruments at the Painter Street Overpass in Rio Dell, which is located within 1.5 km of the bridge, recorded peak ground accelerations of 0.55 g in a direction parallel to the longitudinal direction of the bridge, and 0.39 g in the perpendicular direction, i.e. parallel to the bridge's transverse direction. The bridge performed as anticipated without any damage. In addition, its bearings had completely re-centered by the end of the earthquake, although it was estimated that the bridge had moved 20 cm in its longitudinal direction and 10 cm in its transverse direction.

The performance of the bridge demonstrated the suitability of seismic isolation for retrofitting old bridges that need to be seismically upgraded, and motivated the California Department of Transportation (Caltrans) to initiate a systematic replacement of existing steel bearings with seismic isolation bearings.

5.6.4 The Rangitaiki River Bridge at Te Teko, New Zealand

The Rangitaiki River Bridge at Te Teko, New Zealand, which is part of State Highway 30, crosses the Rangitaiki River. The seismically isolated reinforced-concrete bridge was constructed in 1982. A 6.3 Richter magnitude earthquake, with an epicenter within 14 kilometers of the bridge, occurred on March 2, 1987. The peak ground acceleration recorded nearest to the bridge was 0.33 g, recorded at a distance of 10 km from the bridge (DIS [33]). The bridge has five twenty-meter spans supported by five-meter high piers. Lead rubber bearings were installed at the piers and standard rubber bearings at the abutments.

The overall performance of the isolation system of the bridge was satisfactory. However, the bridge suffered minor damage due to the slippage of one of the two rubber bearings in one of the abutments. The bearing was not properly restrained against sliding and, during the excitation, moved away from its original location. This resulted in small permanent displacement of the superstructure which required temporary disruption of the bridge's operation for repairs. Nevertheless, the performance of the seismic isolation system had been successful. It is estimated that if the bearing had been adequately restrained to avoid slippage, the superstructure would not have experienced any damage (McKay et al. [77]). This incident demonstrated the importance of careful construction and installation of all details to ensure the effectiveness of seismic isolation.

Chapter 6

Practical Considerations for Seismic Isolation

6.1 Seismic Isolation of Buildings

Although this section refers to seismically isolated buildings, many of the issues considered here apply also to seismically isolated bridges. In general, seismic isolation is employed more frequently for bridge structures, probably due to the availability of funds and the consequences to the economy and the public of potential collapse of critical highways during powerful earthquakes.

6.1.1 Configuration of seismic isolation system

Prior to the selection of the plan layout of the isolation devices to avoid torsional effects, a decision must be taken regarding the level where seismic isolation is to be applied. Among the several factors which must be taken into account are the site constraints, the type and nature (new construction or rehabilitation) of the structure, together with ease of inspection, maintenance and replacement of the system in case of malfunction. In addition, the building's internal services, such as elevators and pipes, must be carefully designed incorporating flexible service connections that allow unrestrained relative movement of the superstructure at the isolation level. Finally, it is necessary to have a rigid diaphragm directly above the isolation level to ensure uniform deformations at each isolator.

The two most commonly used seismic isolation systems are shown in Figure 6.1. Both configurations allow continuity of services within a building, such as the operation of elevators and access to stairways. However, flexible service connections should be provided between the public infrastructure services, such as water supply and gas pipes, and building systems in order to allow unrestrained movement of the superstructure relative to the substructure. Similar design considerations are required for entrances, stairs and any other facility that crosses the isolation level.

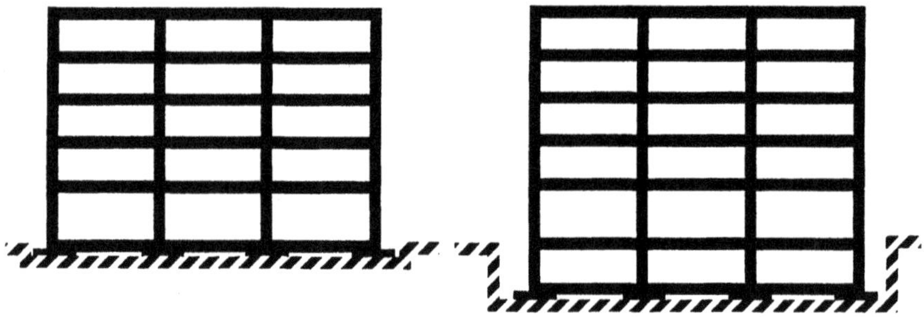

Figure 6.1: Typical seismic isolation configurations.

In addition, a backup system, required as a temporary support for the building, is easily installed. The configuration having the isolation system at basement level requires an additional cost in the case where a basement has not been included in the design. Adopting this configuration involves an additional cost for the required extra retaining wall. Finally, the basement installation requires special care to ensure a dry area for the isolation system. The configuration with the isolation system at ground level can easily provide access for inspection and removal of an isolator.

There are several other, less attractive, configurations for isolation systems involving their insertion at any other level, typically the top of the first story. However, most of these configurations involve additional practical problems and additional cost. One issue is the continuity of services, that of the elevator in particular, which can be dealt with by constructing a cantilevered elevator hanging from the isolated superstructure. In addition, special cladding construction details may be necessary, increasing the cost of seismic isolation. Finally, more sophisticated fire protection of the installed system may also involve additional cost.

6.1.2 Practical issues and problems

A clearance must be ensured for the lifespan of the structure, to accommodate the maximum expected relative displacements at the isolation level, between the isolated superstructure and the foundation level or the non-isolated part of the structure, as shown in Figure 6.2.

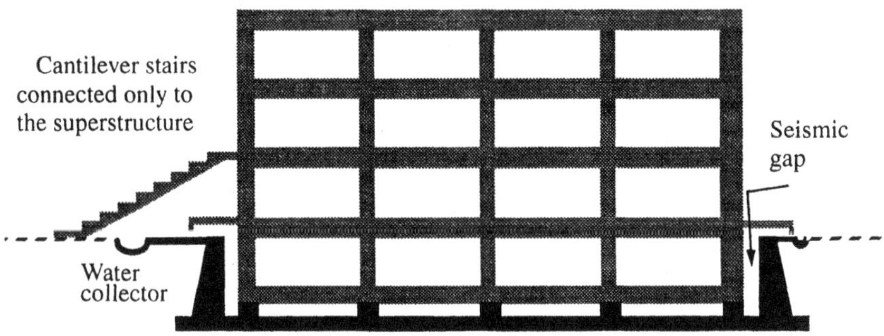

Figure 6.2: Section of a typical seismically isolated building.

When the isolation system is installed below ground level, a moat wall around the building is required to provide this clearance. The isolation gap must be covered to protect both the gap and the isolation system and also as a safety measure against possible accidents due to the presence of this space. Exterior or interior stairs and elevators crossing the isolation level must be cantilevered to allow relative displacement of the superstructure. The utilities and internal services that intersect the isolation level must also allow for the maximum expected displacement. Water, gas, drainage pipes and cable connections must be designed with flexible joints at that level to accommodate such displacement (Figure 6.3).

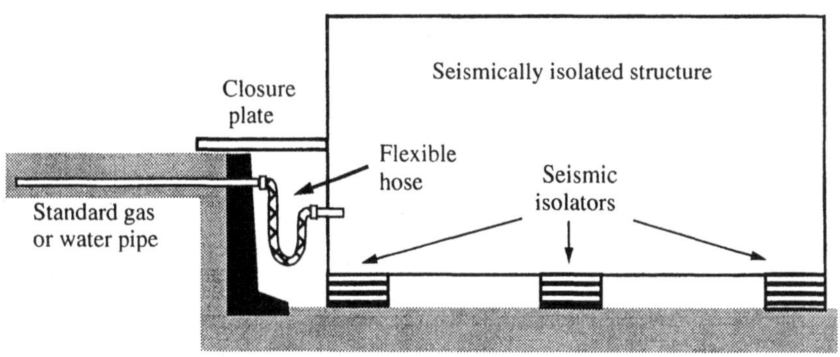

Figure 6.3: Special connectivity details for utilities crossing the isolation level.

No future modification should impose any restriction or obstacle on the superstructure's relative movement. Although there is an additional cost for the flexible joints and the special design and construction details, there are significant advantages due to the substantially reduced inter-story drifts of the superstructure. These allow a less conservative design of the internal utilities which do not have to be able to withstand large deformations. Thus, the possibility of gas leakage, which in some cases may be more destructive that the actual earthquake damage, also decreases. During inspection after a strong earthquake, any leakage at the isolation level due to overstressing of the connections must be detected.

Considering the uncertainties associated with the seismic input and the assumptions and approximations used in the design and analysis, it is reasonable to install backup systems for the case of a higher than expected structural response. Backup systems may be used for both vertical loads and horizontal displacements. They may be used to carry the vertical loads in the case of bearing malfunction, by redistributing the corresponding vertical loads to temporary supports to allow the replacement of the bearing (Figure 6.4) during repairs.

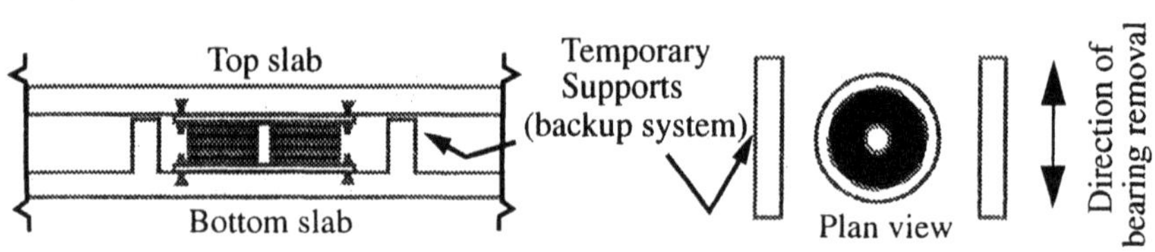

Figure 6.4: Backup system for temporary vertical support.

Backup safety systems and design details that allow and facilitate the removal and replacement of an isolator are highly desirable in case of the failure of an isolator. The temporary supports must be designed to carry vertical loads and transfer them to the foundation. The difficulty involved in the replacement of a bearing depends on the configuration of the isolation system and the location of that bearing. Careful selection of the bearing connections, e.g. dowels, may simplify its removal. Also, design provisions must be made that allow the locating of devices (e.g. jacks) that can be used to raise the structure allowing the removal of damaged isolators.

In addition, constraints as shown in Figure 6.5 and Figure 6.6 may be used as backup systems to restrict the relative displacements at the isolation level within the clearance provided around the structure. This restriction is necessary to avoid collision of the building with adjacent structures or with the moat wall in the case of a more severe earthquake than that anticipated in the design. Such collisions may result in local damage, excitation of higher modes, increased floor accelerations and consequent, building contents damage. Although the constraints will increase floor accelerations and the probability of building contents damage, in the case of an extreme event it is preferable to protect the isolation system and the superstructure from damage due to abrupt shocks. The proposed constraints may be of two types, namely stoppers (Figure 6.5) and bumpers (Figure 6.6).

A backup stopper may be composed of a lead or steel bar, of sufficiently large diameter, surrounded by a cylindrical bearing with an inner diameter slightly less than twice the sum of the radius of the solid bar and the width of the seismic gap. To minimize the shock it is better to design each layer of the laminated cylindrical bearing to have a smaller inner diameter, from bottom to top. Each steel plate and adjacent rubber sheet sticks out slightly from the lower one by a small indent, for the constraint to be applied gradually, instead of abruptly thereby minimizing the resulting mass acceleration increase.

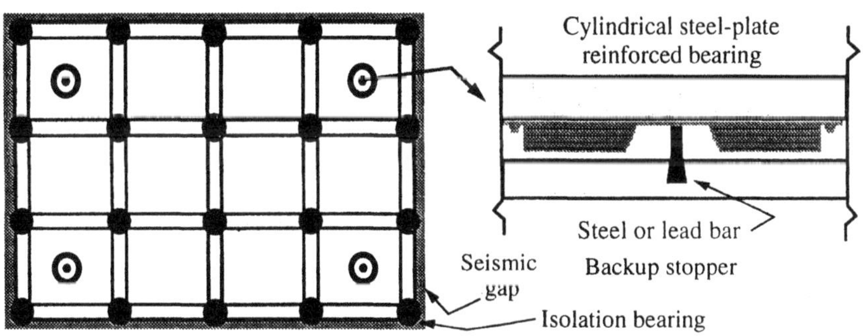

Figure 6.5: Backup stopper system.

As shown in Figure 6.6, pieces of flexible material, e.g. elastomer, preferably with viscous properties, may be installed at particular locations in the perimeter of the building as shock absorbers. They must protrude from the edges of the building in order to be hit first and so to avoid the abrupt direct collision of the structure with the moat wall or adjacent structures. The buffers must be installed at locations where structural members are strong enough to avoid local damage. Additional reinforcement, in the case of reinforced-concrete structures, may increase their strength locally.

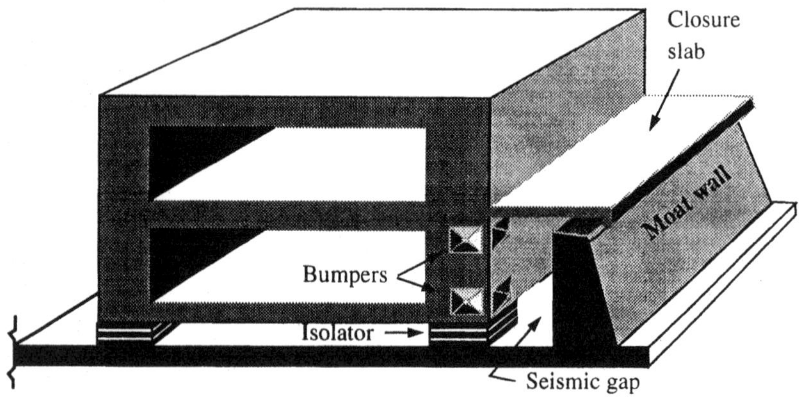

Figure 6.6: Bumpers as shock absorbers.

Another issue of concern is the uplift of the bearings due to overturning moments. The bearings are not designed to carry tensile stresses and therefore such stresses must be avoided or, at least, limited to levels which cannot affect the performance of the bearings. Tensile forces are induced in the bearings when the axial forces due to overturning moments are greater than the forces due to gravity loads. Careful configuration of the bearings and distribution of the gravity loads, which are the more beneficial the higher they are, may eliminate the problem of uplift. Bearings at the edges of shear walls or at corner columns are usually the most vulnerable to uplift. Whenever the tensile forces cannot be reduced to acceptable levels, special connection details may be used to avoid the transmission of tensile forces. In that case, special details must be added next to the bearings to bring them back to their original configuration after uplift movements. Finally, the sequence of the installation of the isolators must be considered, to ensure symmetry and avoid torsional response in the case of an earthquake excitation or any other strong lateral loading during construction.

6.2 Seismic Isolation of Bridges

The recent Northridge and Kobe earthquakes showed the vulnerability of highway bridges and the significant consequences to the public and the economy of their failure. Recent earthquakes have demonstrated the importance of maintaining the operation of bridge structures, especially those of critical highways. Seismic upgrading of critical highway bridges is under way and in many cases seismic isolation is employed to protect them against future strong earthquakes.

In the past, bridge engineers relied on providing excessive stiffness to reduce displacements and deformations during earthquake excitations. However, that design approach was proved problematic since, in many cases, the increased stiffness caused higher seismic forces and resulting damage. For the last few decades, earthquake resistance design for bridge structures has relied mostly on ductility to dissipate the induced seismic energy and prevent structural collapse. Ductility-based design aims at inelastic deformations at certain regions where increased ductility is provided by sufficient shear and confinement reinforcement. Damage of brittle regions and elements, as well as of structural components that are essential for structural stability is prevented by providing increased strength to those regions and elements. However, the ductility-based design approach relies on damage to dissipate the seismic energy, which often causes disruption of the service of the bridge to allow its repair.

Recent years, especially after the latest earthquakes, have seen an increased interest in bridge designs which not only avoid structural collapse, but also prevent excessive damage in order to maintain the continued functioning of the bridge. Many bridge structures must be seismically upgraded, preferably with minimum disruption to their operations. Seismic isolation can be utilized to achieve superior seismic performance, often without significant disruption to the operation of the bridge, since only minimal structural modification is required. It is, in general, more cost effective and practical to upgrade an existing bridge structure than to retrofit an existing building, especially if only the superstructure of the bridge is to be isolated.

6.2.1 Configuration of seismic isolation system

The superstructure of an existing bridge may, in many cases, be seismically isolated by simply replacing the existing common bearings, which often have anyway to be replaced due to deterioration, with isolation bearings. In those cases, seismic isolation is inexpensive and the only additional overhead is the clearance that must be provided for the relative displacements of the superstructure. Significant savings in the cost of piers, abutments and foundations can be achieved in new bridge structure constructions due to the reduced demand for strength and ductility. Similarly, expensive modifications to increase the strength and ductility of these components in retrofitting existing bridges can also be minimized, or even eliminated. Figure 6.7 shows a typical configuration of the isolators to seismically isolate the superstructure of a bridge.

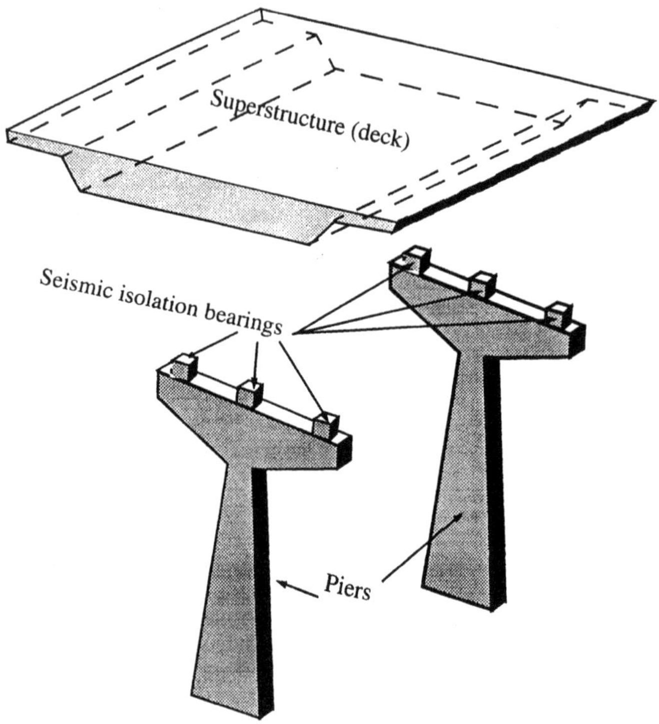

Figure 6.7: Typical configuration to isolate a bridge's superstructure seismically.

By seismically isolating the superstructure of a bridge, the seismic forces induced in the bridge structure and, therefore, its strength and ductility demands as well, are substantially reduced. However, the protection of the piers and abutments, as well as their foundations, is also very important and critical. Although these components are not seismically isolated, they are indirectly protected by seismic isolation since the latter reduces and better distributes the seismic loads that are transferred to them. Proper dimensioning of the isolators can achieve a desirable distribution of the lateral forces according to their strength and ductility capacity. Loads can be directed through substructural elements that have the necessary strength, and away from weaker components of the substructure. In addition, the lateral distribution of the lateral forces can be manipulated, by proper sizing of the isolators, to avoid torsional effects.

Often, the piers have different heights and stiffnesses, due to ground morphologies, that cause undesirable distribution of the lateral loads in the bridge substructure. Consider for example a bridge such as that in Figure 6.8(a) with a pier twice as high as another pier. Under longitudinal seismic motion the short pier will have shear forces about eight times higher than those of the tall pier, assuming that all other characteristics are the same. Therefore, the short pier needs to have greater strength of almost an order of magnitude in order to withstand the large seismic forces. If the earthquake occurs in the transverse direction, in case of eccentric stiffness in that direction, the situation may be much worse due to serious torsional effects.

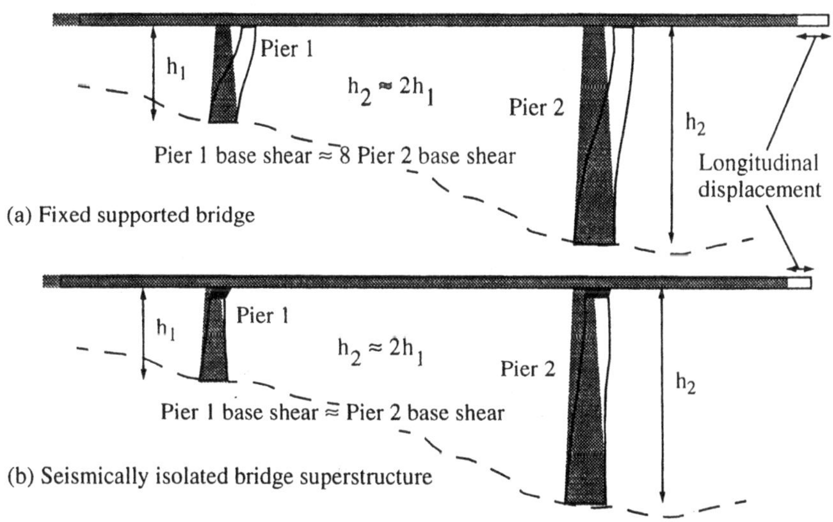

Figure 6.8: Control of shear forces in bridge substructure using seismic isolation.

Seismic isolation of the superstructure and proper dimensioning of the isolators (Figure 6.8(b)) can control the distribution of lateral loads and avoid torsional effects. The displacements of the superstructure are taken mostly at the isolation level by the deformations of the isolators, rather than by the deformations of the piers. In addition, the overall seismic loads that must be transferred to the foundations are significantly reduced and their distribution can be controlled by the characteristics of the isolators.

The shear forces are controlled mostly by the characteristics of the isolators, in particular their stiffness, unless the piers are very flexible. Then, the combined flexibility of the piers and the isolators determines the distribution of the shear forces in the piers. When lead rubber bearings are used, the initial stiffness, i.e. prior to the yielding of the lead plugs, of the isolators determines the distribution of the minor lateral loads, such as wind, minor earthquakes, braking and centrifugal loads. Under a strong earthquake excitation the post-elastic stiffness of the isolators determines the distribution of the shear forces that are transferred to the ground. Therefore, proper design of the isolators can achieve a desirable lateral load distribution and, if necessary, direct loads away from weak bridge substructures.

The seismic elastomeric bearings which use lead plugs or auxiliary lead dampers can function not only as isolators under earthquake excitations but also as bearings for thermal effects. The creep behavior of lead allows it to deform under thermal expansion with substantially smaller forces than those that would have been caused by fast-applied displacements. However, the seismic gaps should be designed for the maximum expected relative displacements, taking into account their anticipated reductions due to temperature changes. In addition, special care should be taken during design and construction, as well as during maintenance, to ensure clearance at the seismic gap. The seismic isolators can be designed to provide the initial rigidity and stiffness under minor loads, such as wind, braking and centrifugal loads.

Seismically isolating the superstructure of a bridge results in piers behaving almost as cantilevers with very limited top constraints. In some cases, especially when the piers are very tall and do not have the strength and capacity to withstand the anticipated seismic loads, it may be preferable to isolate the entire bridge structure. However, this involves much more structural modification and often disruption of the functionality of the bridge.

6.3 Maintenance and Management of the Isolation System

The isolation system must remain operational for the whole expected lifetime of the structure under all possible environmental effects. These may corrode the metallic parts of the isolation system and deteriorate the elastomer; such effects may be reduced by using a protective rubber cover. The maintenance of the isolation system, and especially that of the seismic gap, must be frequent to ensure the proper function of seismic isolation. Inspections must be performed on a regular basis but also after strong seismic events. They must involve examination of metallic parts for corrosion, tightness of bolts, rubber deterioration and any other malfunction. In addition, a water drainage system must be provided to eliminate water as well as trash from the isolation system and thus reduce the possibility of its environmental deterioration.

Inspections are required to ensure that no objects are placed in the gap which might restrict the relative movement of the superstructure and, consequently, the decoupling of the structure from the ground motion.Therefore the residents of isolated buildings must be informed of the main features of the isolation system and the importance of maintenance of the seismic gap. Any arbitrary construction or modification which applies constraints to the relative movement of the superstructure must be prevented.

6.4 Code Requirements

The adoption of well defined and simple-to-apply provisions for seismically isolated structures in the seismic codes, worldwide, is necessary to promote the utilization of this method in the effort to mitigate the consequences of strong earthquake excitations.

In the United States, design codes have been adopted for seismically isolated buildings (UBC [111]), bridges (AASHTO [2]), and hospitals (OSHPD [94]), enabling the wider use of seismic isolation.

6.4.1 Regulations in the United States

In 1980 in the United States, the Structural Engineers Association of Northern California (SEAONC) formed a group to work on design guidelines for seismic isolation. A brief document called *Tentative Seismic Isolation Design Requirements* was published in September 1986, and was later included in the SEAOC's *Blue Book* under the title *Recommended Lateral Design Requirements and Commentary.* In these provisions, emphasis was given to simple equivalent static lateral force procedures.

In 1988, a subcommittee from the SEAOC seismology committee developed a design document called '*General Requirements for the Design and Construction of Seismic Isolated Structures*'. This was included in the fifth edition of the

Blue Book (SEAOC-96 [100]) and later as an appendix to Chapter 23 of the 1991 Version of the Uniform Building Code (UBC) entitled *Earthquake Regulations for Seismic Isolated Structures* (UBC-91 [108]). The latest code emphasizes the use of dynamic analysis, and reduces the cases where equivalent simple static procedures can be used. The 1994 (UBC-94 [110]) and 1997 (UBC-97 [111]) versions of the Uniform Building Code provide an appendix to Chapter 16 giving the regulations for seismically isolated structures.

However, the Uniform Building Code provisions on seismic isolation impose very conservative requirements, which makes the seismic isolation design harder, complicated and costly. Although many new technologies face such conservatism, it may have a negative effect on the wider use of seismic isolation.

Similar provisions were published in 1989 by the California Office of Statewide Health Planning and Development (OSHPD) for application to seismically isolated hospitals, entitled *An Acceptable Procedure For the Design and Renew of California Hospital Buildings Using Base Isolation* (OSHPD-91 [93]). In 1996, the OSHPD of the State of California published the *Division III: Earthquake Regulations for Seismic-Isolated Structures*, (OSHPD-96 [94]), which is similar to the latest Uniform Building Code provisions (UBC-97 [111]).

6.4.1.1 Uniform Building Code (UBC-97 [111])

The objectives of Uniform Building Code (UBC) regulations on seismic isolation and, in general, most other codes related to seismic isolation, are:

• to resist minor and moderate levels of earthquake ground motion without structural or non-structural damage as well as without building contents damage.

• to resist severe earthquakes without failure of the isolation system, with only limited structural and non-structural damage and without the loss of the functionality of the building.

The Uniform Building Code is based on two levels of earthquake motion:

• Design basis earthquake (DBE), which has a 10% chance of being exceeded in a 50-year time period and a corresponding 475-year return period.

• Maximum capable earthquake (MCE), which has a 10% chance of being exceeded in a 100-year time period and a corresponding 1000-year return period.

The isolation system, including utilities crossing the isolation level, must be designed for the maximum capable earthquake and the superstructure must remain 'essentially elastic' for the design basis earthquake.

Uniform Building Code uses two design procedures:

• A static procedure which is based on simple equations to determine the maximum values of displacement and base shear. It is used for low-rise stiff structures, on stiff soils, located away from active faults.

• A dynamic procedure which uses input earthquake excitations and dynamic analysis to determine the maximum values of the response.

The static analysis is required for all seismic isolation designs, although almost all seismic isolation projects use dynamic analyses. Dynamic analysis may be either a time history analysis or a response spectrum analysis.

A response spectrum analysis is required if any of the following is true:

- The building has more than four stories, or 20m in height.
- There are structural irregularities in the horizontal or vertical direction.
- The fundamental period of the fixed-base structure is more than a third of that of the seismically isolated structure under the DBE.
- Site-specific spectra are required, which happens if the structure:
    - is founded on a type S_3 or S_4 soil, i.e. a soft soil,
    - is located within 10 km of a known active fault, or
    - has a fundamental period under the MCE greater than 3 seconds.

Time history analysis, which can be used alternatively to response spectrum analysis, is required if either the isolation system or the superstructure is highly nonlinear.

Static analysis procedure

Although the static analysis method is still based on a single mode of vibration, assuming that the superstructure moves as a rigid body, a large number of coefficients have been introduced. Values for these coefficients can be determined using charts, tables and formulas that are provided in the Appendix to Chapter 16 of the UBC-97 [111].

The total design displacements D_{TD} and D_{TM}, which take into account torsional effects, are estimated by the formulas

$$D_{TD} = D_D\left(1 + y\frac{12e}{b^2 + d^2}\right) \quad \text{in mm,} \tag{6.1}$$

$$D_{TM} = D_M\left(1 + y\frac{12e}{b^2 + d^2}\right) \quad \text{in mm,} \tag{6.2}$$

where e is the actual eccentricity, measured in plan between the center of mass of the superstructure and the center of rigidity of the isolation system, plus 5% accidental eccentricity, in mm; y is the distance, in mm, between the center rigidity of the isolation system and the element of interest in a direction perpendicular to the direction of the seismic loading under consideration; b and d are the shortest and the longest plan size dimensions, respectively, in mm; and, D_D and D_M are the design displacements, in mm, at the center of rigidity of the isolation system in the direction under consideration under DBE and MCE.

The design displacements D_D and D_M are provided by

$$D_D = \frac{C_{VD}T_D(g/4\pi^2)}{B_D} \quad \text{in mm,} \tag{6.3}$$

$$D_M = \frac{C_{VM}T_M(g/4\pi^2)}{B_M} \quad \text{in mm,} \tag{6.4}$$

where g is the acceleration of gravity (9,810 mm/sec^2), B_D and B_M are the damping coefficients, C_{VD} and C_{VM} are the spectral seismic coefficients, and T_D and T_M are the fundamental periods of the isolated structure corresponding to the DBE and MCE levels.

Figure 6.9 shows the design and total displacements.

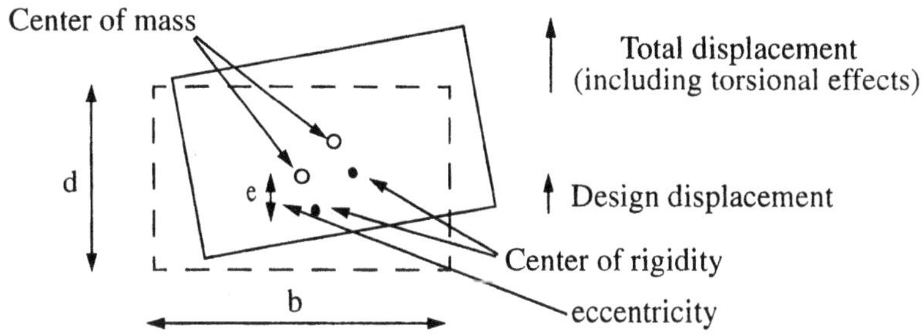

Figure 6.9: Design and total displacements.

The values of the damping coefficients B_D and B_M are provided in tabular form (UBC-97 [111], Table A-16-C) in terms of the effective damping coefficients β_D and β_M, respectively. The latter correspond to the design and maximum displacements, respectively, and can be computed from the following formulas:

$$\beta_D = \frac{1}{2\pi}\left(\frac{total\ area\ of\ hysteresis\ loop}{D_D^2 K_{D,max}}\right),$$

$$\beta_M = \frac{1}{2\pi}\left(\frac{total\ area\ of\ hysteresis\ loop}{D_M^2 K_{M,max}}\right).$$

The terms $K_{D,max}$ and $K_{M,max}$ are equal to the maximum values of $K_{D,eff}$ and $K_{M,eff}$, respectively, as determined by testing.

$$K_{D,eff} = \frac{F_D^+ - F_D^-}{D_D^+ - D_D^-} \quad in\ kN/mm,$$

$$K_{M,eff} = \frac{F_M^+ - F_M^-}{D_M^+ - D_M^-} \quad in\ kN/mm,$$

where F_D^+, F_D^-, F_M^+, F_M^-, D_D^+, D_D^-, D_M^+ and D_M^- are the maximum and minimum, forces and displacements corresponding to the DBE and MCE. The 'total area of hysteresis loop' is the energy that is estimated to be dissipated per cycle of loading of isolator prototypes.

The periods T_D and T_M which correspond to the DBE and MCE are computed using the following formulas, respectively:

$$T_D = 2\pi\sqrt{\frac{W}{gK_{D,min}}} \quad \textit{in seconds,}$$

$$T_M = 2\pi\sqrt{\frac{W}{gK_{M,min}}} \quad \textit{in seconds,}$$

where W is the weight of the building, and the terms $K_{D,\,min}$ and $K_{M,min}$ are equal to the minimum values of the effective stiffness $K_{D,eff}$ and $K_{M,eff}$, respectively, as determined by testing.

The spectral seismic coefficients (C_{VD}, C_{VM}, C_{AD} and C_{AM}) are used to determine the minimum spectral ordinates that should be used in the design. Coefficients C_{VD} and C_{VM} correspond to the constant-velocity regions of the DBE and the MCE, respectively. Similarly, coefficients C_{AD} and C_{AM} correspond to the constant-acceleration regions of the DBE and the MCE, respectively. The values of the coefficients C_{VD} and C_{AD} are provided in charts as functions of the soil profile type (S_A, S_B, S_C, S_D or S_E) and the seismic zone factor (*1, 2A, 2B, 3 or 4*). For seismic zone factor 4 the provided values for the spectral seismic coefficients C_{VD} and C_{AD} must be multiplied by the corresponding near source factors N_v and N_a, respectively. Actually, the coefficients C_{VD} and C_{AD} are the same as the C_V and C_A coefficients that are used in conventional earthquake design (Tables 16-Q and 16-R, UBC-97 [111]). The values of the coefficients C_{VM} and C_{AM} are also provided in charts as functions of the soil profile type (S_A, S_B, S_C, S_D or S_E) and the products $M_M Z N_v$ and $M_M Z N_a$, respectively, where Z is the seismic zone factor, N_v is the near source factor, and M_M is the response coefficient for the MCE.

The value of the seismic zone factor, Z, which ranges from 0.075 to 0.40, can be determined from a provided chart based on the seismic zone factor, which has values of *1, 2A, 2B, 3* or *4* according to a provided US map (Figure 16-2, UBC-97 [111]).

The site soil profile type (S_A, S_B, S_C, S_D or S_E) is provided in a table (Table 16-J, UBC-97 [111]) based on the average shear wave velocity in the top 30.5 m of soil. The coefficient for hard rock is S_A and for soft soils is S_E.

The seismic source type coefficient is determined by the magnitude of the earthquake that can be produced by the adjacent seismic fault. For magnitudes higher than 7.0 the source type is *A*, for less than 6.5 it is *C*, and for all others it is *B*.

Near source effects are taken into account using the coefficients N_a and N_v, which are provided in tabular form (Table 16-S and Table 16-T, respectively, UBC-97 [111]) based on the seismic source type (*A*, *B* or *C*) and the closest distance to known seismic sources

The maximum capable earthquake (MCE) response coefficient M_M is determined from a table (Table A-16-D, UBC-97 [111]) based on the value of the product of the seismic zone factor Z and the near source factor N_v. The MCE response coefficient M_M is used to estimate the MCE response.

The base shear force that should be used for the design of the structural elements below the isolation level is computed using the formula

$$V_b = K_{D,max} D_D \qquad in\ kN. \tag{6.5}$$

The base shear that should be used for the design of the structural elements above the isolation level is given by the formula

$$V_s = \frac{K_{D,max} D_D}{R_I} \qquad in\ kN, \tag{6.6}$$

where R_I is a ductility factor (design force reduction factor) that ranges from 1.4 to 2.0 (Table A-16-E, UBC-97 [111]), and is used to account for the ductility capacity of the superstructure of the seismically isolated structure.

The base shear V_s should never be less than

- the base shear that corresponds to the wind load;
- the seismic forces required for the corresponding fixed-base structure;
- 1.5 times the lateral force that is required fully to activate the isolation system, e.g. the yield load of the LRBs.

Although the expected vertical distribution of the forces on a seismically isolated structure tend to be uniform, a triangular distribution is used to account for higher mode contributions. The lateral force F_x at level x is computed using the following formula from the base shear V_s:

$$F_x = V_s \frac{h_x w_x}{\sum_{i=1}^{N} h_i w_i} \tag{6.7}$$

The maximum inter-story drifts should not exceed $0.01/R_I$.

Dynamic analysis procedure

The dynamic procedure can be either a linear response spectrum method or a time history analysis. The dynamic analysis may give design displacements and forces which are less than those obtained from static analysis. However, the total design displacement D_{TD} and the total maximum displacement D_{TM} for the isolation system cannot be reduced to less than 90% and 80% of those obtained using the static procedure, D_D and D_M, respectively.

The code permits a further reduction by using D_D' and D_M', instead of D_D and D_M, in the formulas of static analysis, where the former are computed by the following formulas:

$$D_D' = \frac{D_D}{\sqrt{1 + (T/T_D)^2}},$$

$$D_M' = \frac{D_D}{\sqrt{1 + (T/T_M)^2}},$$

where T is the elastic fixed-base period of the structure above the isolation level, as determined by Section 1630 of the UBC, entitled *"Minimum Design Lateral Forces and Related Effects"* (Section 1630, UBC-97 [111]).

The design lateral shear force on the isolation system and the structural elements below it shall not be taken as less than 90% of the base shear force computed using the static procedure, i.e. the value computed by Equation (6.5).

The design lateral shear force on the structure above the isolation system shall not be taken as less than 80% of the base shear V_s as computed in the static procedure by Equation (6.6) or the associated specified limits. If the structure is irregular in configuration, the design lateral shear force on the structure above the isolation system shall not be taken as less than the base shear V_s, as computed in the static procedure by Equation (6.6) or the associated specified limits. If dynamic time-history analysis is used, the limits for the above two cases are 60% and 80%, for structures regular and irregular in configuration, respectively.

The response spectrum analysis is sufficient for dynamic analysis if the structure is located on soil profile type S_A, S_B, S_C *or* S_D, and the isolation system has restoring force capabilities, has force deflection properties that are independent of the rate of loading or vertical and bilateral load, has effective stiffness at the design displacement that is greater than one-third of the effective stiffness at 20% of the design displacement, and does not limit maximum capable earthquake displacement to less than C_{VM}/C_{VD} times the total design displacement. Otherwise, a dynamic time history analysis should be performed.

For a response spectrum analysis, site-specific spectra are required for all structures with an isolated period, T_M, greater than 3.0 seconds, or which are located on soil profile type S_E or S_F, or which are located within 10 km of an active fault, or which are located in seismic zone 1, 2A or 2B. Structures that do not require site-specific spectra can be designed using the design response spectra provided by the Uniform Building Code (Figure 16-3, Chapter 16, Division III, UBC-97, [111]). In any case, the design spectrum shall not be taken as less than the Design Response Spectra provided by the UBC.

The values of the seismic coefficients C_a and C_v shall be taken equal to the values of C_{AD} and C_{VD}, respectively, to construct the design spectrum for the design-basis earthquake. Similarly, for the response spectrum for the maximum capable earthquake the values of C_{AM} and C_{VM} should be used for the seismic coefficients C_a and C_v, respectively. This spectrum shall be used to determine the maximum displacement and overturning forces for design and testing of the isolation system. If a site-specific spectrum is determined for the maximum capable earthquake, then the design spectrum shall not be taken as less than 80% of the design response spectra provided by the Uniform Building Code using the values of C_{AD} and C_{VD} for the seismic coefficients C_a and C_v, respectively.

For a time history analysis, at least three pairs of horizontal components from recorded seismic events that are representative of the site, soil and earthquake characteristics must be used. In addition, when the structure is either founded on soft soil or located very close to an active fault, a site-specific ground motion must be used in the design. For each pair of horizontal ground-motion components, the square root sums of the squares (SRSS) of the 5% damped spectrum are computed. The ground motions are then scaled so that the average of the SRSS spectra does not fall below 1.3 times the 5% damped spectrum of the design-basis, or maximum capable earthquake, by more than 10% over $0.5T_D$ to $1.25T_M$ seconds. The design must be based on maximum response quantities if three time histories are used, and can be based on average response quantities if seven time histories are used.

The isolation system should be modeled using deformational characteristics developed and tested according to certain requirements and tests performed in accordance with a certain section of the Uniform Building Code (Section 1665, UBC-97, [111]). The modeling of the isolation system should account for the spatial distribution of the isolators, assess overturning/uplift forces on individual isolators, account for the effects of vertical, bilateral and/or the rate of loading whenever the force-deflection properties of the isolation system are dependent on any of them, and calculate translation in both horizontal directions and torsion of the superstructure considering the most disadvantageous location of mass eccentricity.

The design-basis earthquake shall be used to calculate the total design displacement of the isolation system and the lateral forces and displacements of the isolated structure. The maximum capable earthquake shall be used to calculate the total maximum displacement of the isolation system.

For a response spectrum analysis, the total design and maximum displacements shall include simultaneous excitations of the model by 100% of the most critical direction of ground motion and 30% of the ground motion on the orthogonal axis. The maximum displacement of the isolation system shall be calculated as the vectorial sum of the two orthogonal displacements. However, the 30% increase may be overconservative, since the isolation systems are, in general, isotropic and the maximum displacement can be in any direction.

For a time history analysis each pair of ground motions shall be applied simultaneously to the model considering the most disadvantageous location of the mass eccentricity. The maximum displacement of the isolation system shall be calculated as the vectorial sum of the two orthogonal displacements at each time step. The parameter of interest shall be calculated for each time history analysis. If three time histories are used, the maximum response of the parameter of interest shall be used, while if seven time histories are used, its average value may be used in the design.

The isolation system, foundation and structural elements below the isolation system shall be designed using the appropriate provisions of non-isolated structures. The structural elements above the isolation system shall be designed using the appropriate provisions of non-isolated structures, and the forces obtained from the dynamic analysis divided by the ductility factor R_I.

The maximum inter-story drift ratio of the structure above the isolation system, when calculated by response spectrum analysis, shall not exceed $0.015/R_I$, while when calculated by time history analysis it shall not exceed $0.02/R_I$. If the inter-story drift ration exceeds $0.01/R_I$, the secondary effects of maximum capable earthquake lateral displacement Δ shall be investigated.

General requirements

Elements of seismic isolated structures, which are at, or above, the isolation interface shall be designed to resist a total lateral seismic force equal to the maximum dynamic response of the element under consideration. Elements that cross the isolation interface shall be designed to withstand the total maximum displacement. Finally, elements below the isolation interface shall be designed according to the requirements of the corresponding section of the code for non-isolated structures (Section 1632, UBC-97, [111]).

Global overturning of the structure at the MCE with the full dead load being used should be considered and checked. When some isolators uplift, others experience increased vertical loads for which the isolators are required to be designed and tested. The design should ensure that an isolator remains stable when displaced to the total maximum displacement D_{TM} under $1.2DL + 1.0LL + E_{max}$ and $0.8DL - E_{min}$ at the MCE, where E_{min} and E_{max} are the minimum downward and maximum upward, respectively, vertical loads on an isolator due to superstructure overturning. The effects of vertical accelerations are taken into account by the increased or decreased dead load.

Detailed requirements for the isolation system, in particular concerning environmental conditions, wind forces, fire resistance, restoring force, overturning, vertical load stability, inspection and replacement, and for the structural system are provided by the Uniform Building Code (Section 1661, UBC-97, [111]).

In addition, an independent engineering team must verify the validity of the design and construction procedure of a seismically isolated structure (Section 1664, UBC-97, [111]). The review team should evaluate the seismic criteria and the design, both preliminary and final, that has been used.

Finally, certain specifications for the required tests of isolation systems are provided by the Code (Section 1665, UBC-97, [111]).

6.4.1.2 Codes for seismic isolation of hospitals

The California Office of Statewide Health Planning and Development (OSHPD) code for seismically isolated hospitals, which is entitled *Division III: Earthquake-Regulations for Seismic-Isolated Structures* (OSHPD-96 [94]), is similar to the 1997 UBC ([111]). However, site-specific ground motions, which must be evaluated and approved by the California Department of Mines and Geology (DMG) on behalf of the OSHPD, are required. Both static and dynamic analyses must be carried out. Almost always, the dynamic analysis must include a time history analysis. Much more restrictive requirements are specified for the design review.

Access for monitoring and replacement of the isolation system must be provided for the entire life of the seismically isolated building. The latter must be instrumented with accelerometers. After every strong earthquake the isolated structure, the isolators and the accelerometers must be inspected. Finally, some isolators must be removed every ten years to be tested.

6.4.1.3 Codes for seismic isolation of existing buildings

Both UBC and OSHPD codes regulate the design of new structures, without specifically covering retrofitting of existing ones. However, retrofit projects use both codes, as well as the *Guidelines for Seismic Rehabilitation of Buildings* (FEMA [36]) and its commentary (FEMA [37]).

6.4.1.4 Codes for seismic isolation of bridges

In the United States, after the 1971 San Fernando earthquake, the Federal Highway Administration (FHA) and the Applied Technology Council (ATC) recommended bridge-design specifications which were published in the report FHWA/RD-81/081, known as ATC-6. These specifications were adopted by the American Association of State Highway and Transportation Officials (AASHTO) as specifications for highway bridges.

The initial AASHTO *Guide Specifications for Seismic Isolation Design* ([1]), developed in 1991, focused on elastomeric bearings, which were the only reliable isolation bearings at that time in the United States. The revised *Guide Specifications for Seismic Isolation Design* ([2]), approved by AASHTO in 1998, provides recommendations for sliding isolation bearings as well as for elastomeric bearings.

6.4.2 Regulations in Japan

In Japan, there are several guidelines for designing and constructing seismically isolated structures. Although seismic isolation is widely used in Japan, no established codes exist and the construction of a seismically isolated structure requires special permission from the Minister of Construction. The approval of an application and the provision of the relevant permission is assigned to the Building Center of Japan (BCJ) (Kani and Manabu [54]). The application is evaluated by the Committee for Evaluation of Base-Isolated Structures.

In 1987 a base-isolated structures subcommittee of the Architectural Institute of Japan (AIJ) published a set of guidelines entitled *Recommendation for the design of Base Isolated Buildings*, which presents some design guidelines for seismically isolated structures. This document, which was revised in 1993, includes design guidelines, commentary, case studies and design information which are useful and informative, but still of no legal standing.

In addition, the Building Center of Japan has published a report, *Report on safety Evaluation Guideline for Base-Isolated Structures*, that provides guidelines for designing and constructing seismically isolated buildings. Design guidelines for seismic isolation of bridges are provided in the *Specifications for Highway Bridges* established by the Japan Highway Public Corporation. Finally, design guidelines for nuclear power facilities have been established by the Central Research Institute of the Electric Power Industry (CRIEPI).

Chapter 7

Structural Modeling and Analysis of Seismically Isolated Structures

7.1 Modeling of Response

The complexity of the model to be used in the various stages of design and analysis of a seismically isolated structure depends on the purpose of the model at a particular stage, and the desired accuracy of the results. The simplest model is a single degree-of-freedom (DOF) system, which assumes that the superstructure can be represented as a single mass and that the stiffness is equal to that of the isolation system. This model, shown in Figure 7.1(a), approximates sufficiently well the overall behavior of the isolated structure at the preliminary design stage. The response of a seismically isolated system is much more significantly dominated by its fundamental eigenmode than the response of the same system when fixed supported. Thus, single DOF modeling is actually more appropriate for a seismically isolated structure than for a fixed supported one. This may be demonstrated by examining the eigenmodes of a structure for both cases of boundary conditions.

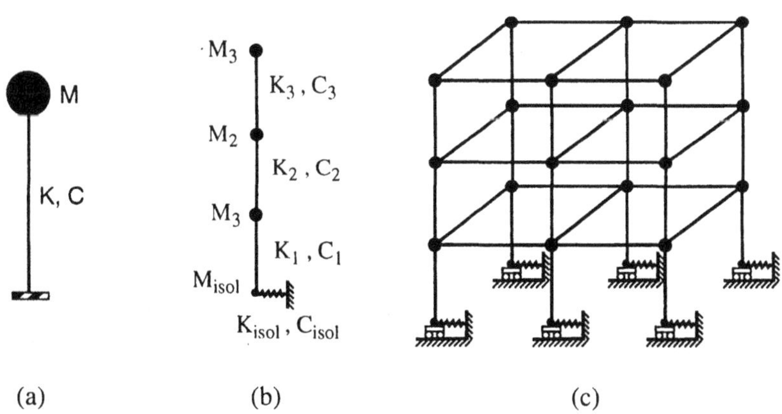

Figure 7.1: Models for the analysis of isolated structures.

A finer model is the "shear building" shown in Figure 7.1(b). It assumes that the overall deformation of the building may be modeled as a shear beam and that the mass of the building is lumped at the floor levels. The model is a lumped mass multi-DOF system and assumes that the beams of the structure are sufficiently rigid to prevent rotations of the masses at the floor levels about the horizontal axis. In general, it is an appropriate model for relatively stiff buildings, i.e. low- to medium-rise, which are also those for which seismic isolation is suitable. The formulation of the required matrices and the methodology for the numerical computation of the dynamic response for such a model are described in Appendices A and B. Input earthquake excitation is applied only in one direction, since each mass has only one horizontal dynamic degree of freedom.

A three dimensional (3-D) model with the masses lumped at the nodes, as shown in Figure 7.1(c), is required to compute detailed local response effects and quantities, such as vertical displacements, torsional effects due to eccentricities, and the horizontal distribution of seismic forces. The superstructure is modeled as a 3-D frame with the mass concentrated at the nodes. The high planar rigidity of the slab of each floor allows constraint of the corresponding degrees of freedom of each node which lies at the level of the slab, thus reducing the required computational work.

In addition to the appropriate choice of a model for the seismically isolated structure, another issue of concern, either at the preliminary design stage or in the final analysis, is the nonlinear behavior of some isolation systems. The nonlinear behavior at the isolation level may be approximately linearized. Linearizations are reasonable for some isolation systems, especially when considering the inherent uncertainties involved in the dynamic response due to an earthquake excitation. Note that some isolation systems are characterized by highly nonlinear behavior and hysteretic dissipation of energy through cycles of loading and unloading which cannot be neglected, at least at the final analysis stage. All the above presented models may be used to consider such local non-linearities; the procedures used to account for the non-linearities for both the single-DOF system and the multi-DOF shear beam system are described in Appendix B.

7.2 Natural Modes of Vibration

The response of a linear-elastic structure with N degrees of freedom may be computed through modal analysis as the summation of the responses of N independent single DOF systems corresponding to the N normal modes of vibration of the system. The participation factors of each mode indicate the importance of the fundamental eigenmode and the actually rigid body motion of the superstructure.

In addition, insight into the response of an isolated building can be gained by considering that the eigenmodes are the deformed shapes of the oscillating building. Finally, the single-DOF approximation in the preliminary design may be verified by the high contribution of the first mode. Modal analysis and the mode superposition method are used extensively to perform linear elastic analysis of multi-DOF structures; the system of differential equations of motion may be uncoupled as a set of single-DOF differential equations, for which the response is usually available in the form of a response spectrum.

The system of equations of motion for a lumped multi-DOF system, such as that in Figure 7.1(b), excited by an earthquake is:

$$\lfloor M \rfloor \lfloor \ddot{U}(t) \rfloor + \lfloor C \rfloor \lfloor \dot{U}(t) \rfloor + \lfloor K \rfloor \lfloor U(t) \rfloor = -\lfloor M \rfloor \lfloor \ddot{U}_g(t) \rfloor \tag{7.1}$$

When the system is linear elastic, mode superposition is appropriate. Using this method (Appendix A) the obtained independent equations, which correspond to the eigenmodes of the structure, are of the form

$$\ddot{\psi}_i + 2\,\xi_i\,\omega_i\,\dot{\psi}_i + \omega_i^2\,\psi_i = -\ddot{u}_g \tag{7.2}$$

where i is the index of a particular mode.

The displacements and accelerations of the masses of the structure may be computed using mode superposition:

$$\lfloor U(t) \rfloor = \lfloor \Phi \rfloor \lfloor y(t) \rfloor = \sum_{i=1}^{N} \lfloor \phi_i \rfloor\, y_i(t) = \sum_{i=1}^{N} \lfloor \phi_i \rfloor\, \Gamma_i\, \psi_i(t) \tag{7.3}$$

$$\lfloor \ddot{U}_{total}(t) \rfloor = \lfloor \ddot{U}(t) \rfloor + \lfloor \ddot{U}_g(t) \rfloor = \sum_{i=1}^{N} \lfloor \phi_i \rfloor\, \Gamma_i\, \ddot{\psi}_i(t) + \lfloor \ddot{U}_g(t) \rfloor \tag{7.4}$$

The participation factor Γ_i for each mode i is defined as:

$$\Gamma_i = \frac{\lfloor \phi_i \rfloor^T \lfloor M \rfloor \lfloor e \rfloor}{\lfloor \phi_i \rfloor^T \lfloor M \rfloor \lfloor \phi_i \rfloor} = \frac{L_i}{m_i}$$

The effective modal mass M_i^* is the percentage of mass oscillating with the ith mode, and is expressed by the following expression:

$$M_i^* = \frac{L_i^2}{m_i} = \frac{\left([\phi_i]^T [M] [e]\right)^2}{[\phi_i]^T [M] [\phi_i]}, \quad M_{total} = \sum M_i^* \tag{7.5}$$

Modal analysis shows the higher contribution of the lower eigenmodes, the fundamental mode in particular; higher modes may therefore be neglected, at least in the preliminary design. This is evident from Figure 7.2, which shows the variation of the effective modal mass M_i^* for the first two modes, expressed as a percentage of the total mass M_{total}, as the degree of isolation changes. The latter is expressed as the ratio of the fundamental mode of the isolated structure to the fixed supported superstructure.

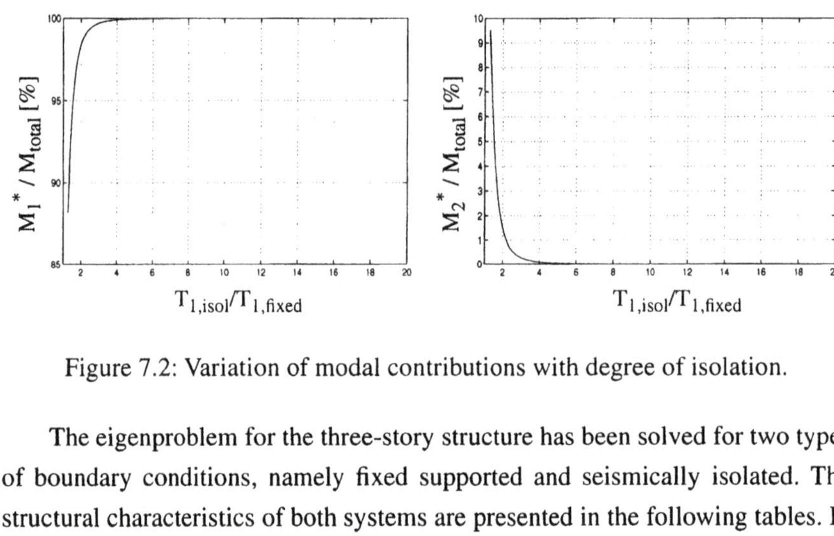

Figure 7.2: Variation of modal contributions with degree of isolation.

The eigenproblem for the three-story structure has been solved for two types of boundary conditions, namely fixed supported and seismically isolated. The structural characteristics of both systems are presented in the following tables. In particular, Table 7.1 presents the characteristics of the fixed-supported system, which are the same as those of the fixed-supported system presented in Chapter 3. The characteristics of the seismically isolated system are presented in Table 7.2. Note that a stiffness of 50 MN/m is used to represent the stiffness of the isolation value. This is a typical value for a linear seismic isolation system. A smaller value, in particular 30 MN/m, has been used in an example in Chapter 3, as it lies between the stiffness of the linear (50 MN/m), and the post-yield stiffness of the bilinear isolation system (i.e. 25 MN/m) that are used in this chapter.

Level	Masses	Stiffness
	[tons]	[MN/m]
1	305	1250
2	285	1125
3	265	1125

Table 7.1: Structural characteristics of a fixed-supported system.

Level	Masses	Stiffness
	[tons]	[MN/m]
Isolation	335	50
1	305	1250
2	285	1125
3	265	1125

Table 7.2: Structural characteristics of a seismically isolated system.

The results of the eigenproblem analysis are summarized in Table 7.3 for the fixed supported system, and in Table 7.4 for the isolated system. The eigenmodes of the two systems are shown in Figure 7.3 and Figure 7.4, respectively.

Normal mode	Period [sec]	Frequency [rad/sec]	Mode shapes ([M] normalized) mass displacement x 10000			Effective mass[tons]	% of total mass
1	0.21	29.22	6.00	11.28	14.11	771.5	90.23
2	0.08	79.35	-13.65	5.52	-11.41	73.40	8.58
3	0.06	113.03	10.28	-13.90	6.92	10.1	1.18

Table 7.3: Natural characteristics and effective modal mass: fixed supported system.

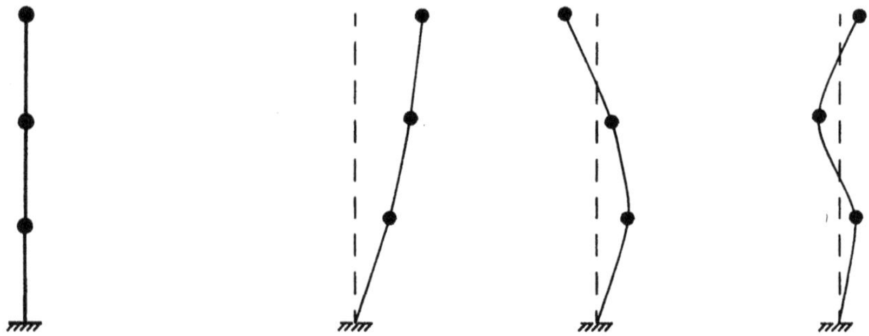

Natural mode shapes: 1st 2nd 3rd

Figure 7.3: Natural mode shapes of the fixed supported building.

The first and usually most essential mode of the isolated structure is characterized by the rigid body motion of the superstructure, as shown in Figure 7.4, and the deformation of the isolation system. Inertia forces due to the seismic excitation are uniformly distributed which in turn implies lower overturning moments. In contrast, in the case of a fixed supported structure, the distribution of these forces increases linearly with height. The fundamental eigenmode is essentially independent of the stiffness and damping properties of the superstructure, considering the actually rigid body motion of the latter due to the large difference between its stiffness and that of the isolation system. Therefore, to a first approximation, a seismically isolated structure may be represented by a single DOF model with the mass of the whole structure above the isolation level and the stiffness and damping properties of the isolation system.

Normal mode	Period [sec]	Frequency [rad/sec]	Mode shapes ([M] normalized) mass displacement x 10000				Effective mass[tons]	% of total mass
1	0.96	6.38	8.87	9.13	9.31	9.41	1189.7783	99.929
2	0.13	48.12	11.25	4.64	-5.65	-1.26	0.2062	0.068
3	0.07	89.61	8.54	-9.52	-8.82	9.86	0.0101	0.003
4	0.05	115.66	-4.52	11.50	-12.43	5.77	0.0010	0.0002

Table 7.4: Natural characteristics and effective modal mass: isolated system.

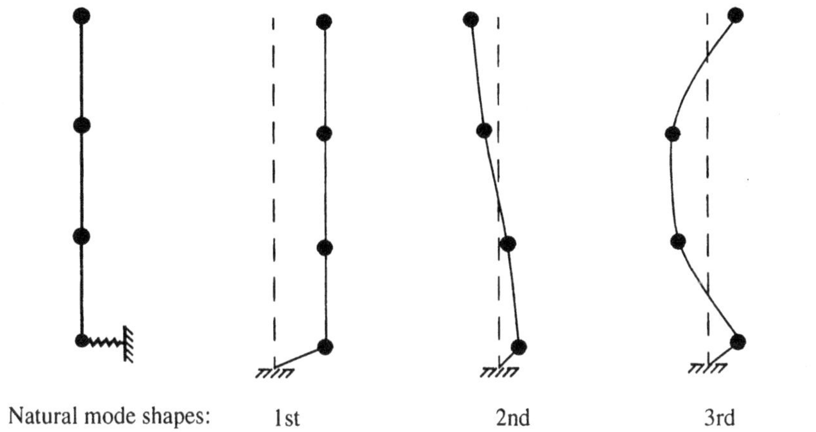

Figure 7.4: Natural mode shapes of the seismically (linearly) isolated building.

Although the fundamental mode dominates the response of a seismically isolated structure, the participation of higher modes increases with the viscous damping of the isolation system. Considering the shape of the displacement response spectrum (Figure 2.3), the displacements, and those of the isolated system in particular, are completely dominated by the fundamental mode. However, the acceleration response spectrum (Figure 2.4) indicates that higher modes have higher spectral values than the fundamental mode of the isolated system, and thus their contribution may be increasingly significant. Despite this fact, however, the small participation factors of these higher modes suppress their potential importance.

7.3 Linear Systems

Some isolation systems may be considered linear-elastic with viscous damping properties, while others are inherently non-linear inelastic with a hysteretic energy dissipation mechanism. The most commonly used linear isolation systems are the natural rubber bearings (NRBs) combined with an auxiliary viscous damper, and the high damping rubber bearings (HDRBs). Although the latter involve hysteretic dissipation of energy and nonlinear behavior, they may still be treated as linear, using equivalent stiffness and damping properties. Linear systems have a constant, low isolation stiffness (Figure 7.5(a)) which makes them vulnerable to dynamic loadings with dominant frequencies close to the fundamental frequency of the isolated structure. The viscous energy dissipation mechanism of linear systems is constant and continuously available for both low and large strains. The differential equation of motion of a linear, viscously damped single-DOF system is

$$M\ \ddot{U} + C\ \dot{U} + K\ U = -m\ \ddot{U}_g$$

which can alternatively be written as

$$\ddot{U} + 2\ \xi\ \omega\ \dot{U} + \omega^2\ U = -\ddot{U}_g$$

with $\xi = \dfrac{C}{C_{cr}} = \dfrac{C}{2\ M\ \omega}$ the damping coefficient,

$\omega = \sqrt{\dfrac{K}{M}}$ the frequency of an undamped SDOF system, and

$\omega_d = \sqrt{\left(\dfrac{K}{M}\right)^2 - \left(\dfrac{C}{2\ M}\right)^2} = \omega\ \sqrt{1 - \xi^2}$ that of an damped SDOF system.

The shear stiffness of an elastomeric bearing may be computed by the following formula, which considers only shear deformations of the rubber and neglects bending deformations as well as the influence of axial (normal) forces. The influence of the axial load is negligible when the axial stresses are relatively low, e.g. well below the critical-for-buckling value. Shear modulus of typical elastomeric bearings are of the order of 0.7 to 1.0 MPa.

$$K = \frac{G_r\ A_r}{h_r} \tag{7.6}$$

The above formula is a very rough approximation of the horizontal stiffness that can be used only for such simple analyses, and for preliminary design purposes. For the final design and analysis phases a more rigorous model needs to be employed to take into account other factors, such as the dependency of the horizontal stiffness on shear, as well as vertical, stresses and strains.

The viscously dissipated energy depends on the velocity of the mass and is characterized by smooth elliptical hysteresis loops, as depicted in Figure 7.5(b).

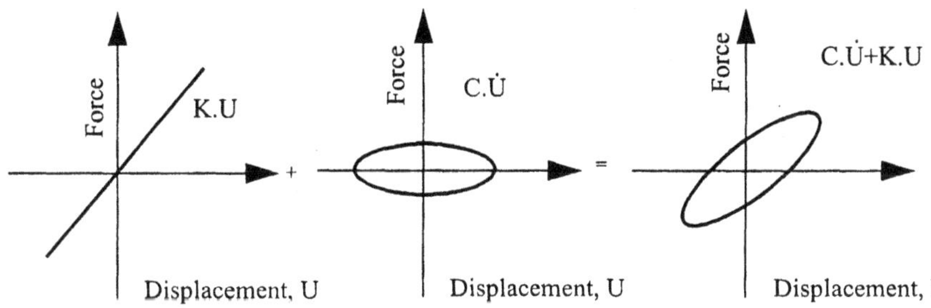

(a) Deformation forces (b) Damping forces (c) Deformation and damping forces

Figure 7.5: Structural characteristics of a linearly isolated system.

As damping increases, the displacements of the structure are reduced; the accelerations decrease until the damping ratio reaches approximately 20-30%, and then increase only slightly for higher damping ratios. The occurrence of this phenomenon is demonstrated here through a dynamic time history analysis of a linearly isolated four-mass model. The structural characteristics of the model are shown in Figure 7.6(a). The introduced damping ranges from 2 to 60% for the fundamental mode, and 2% for the other modes. The dynamic analysis is performed for two ground accelograms, the El-Centro and Olympia earthquakes, scaled to have peak ground acceleration equal to 0.50 g. The maximum relative displacement (Figure 7.6(b)) and the normalized mass acceleration (Figure 7.6(c)) of the structure are plotted as a function of the varying damping ratio.

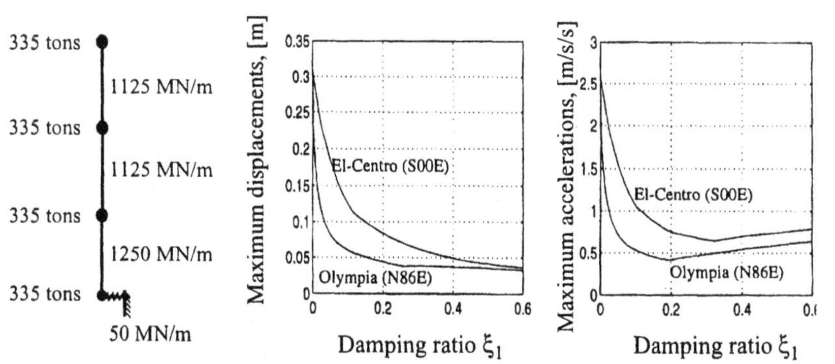

(a) Structural characteristics (b) Maximum displacements (c) Norm. Max. accelerations

Figure 7.6: Effect of viscous damping on a linear isolation system.

7.4 Bilinear Systems

The most commonly used isolation systems that exhibit bilinear behavior are the lead rubber bearings (LRBs) and the natural rubber bearings (NRBs) combined with an auxiliary lead or steel damper. Bilinear systems have a high initial rigidity which is very useful for service lateral loads.

Bilinearly isolated structures typically have a more efficient and reliable energy dissipation mechanism. However, this mechanism is not always available since it is incorporated only when a yield force is exceeded. In addition, highly nonlinear isolation systems may result in higher mode effects which may, in turn, significantly increase the masses' accelerations and consequently the inertia forces induced in the structure and its contents. This situation is worth considering when the prevention of building contents damage is the dominant design requirement. The contribution of higher modes increases with the sudden changes of the isolation system stiffness during reversed loading. The procedure used to perform a step-by-step time history analysis considering the local nonlinearities is described in Appendix B.

7.5 Comparative Study of a Linearly and a Bilinearly Isolated Structure

The response of a typical three-story structure, mounted on linear and bilinear isolation systems, and a hybrid isolation system, is investigated. Linear isolation is typically implemented using high-damping rubber bearings. The most commonly used bilinear system is one containing lead rubber bearings. A linearly isolated structure has higher displacements than a bilinear one because the response of the latter is dominated by the prior-to-yielding high rigidity. This generates higher mode effects due to the coincidence of the fundamental frequency of the initially stiff isolated structure and the superstructure. A hybrid isolation system, combining high-damping rubber with lead plugs (LHDRBs), is an alternative and more efficient system. The characteristics of this system have been discussed in detail in Chapter 4.

Tables 7.3-7.5 summarize the characteristics of a structure which has been isolated with a linear, a bilinear and a hybrid isolation system, respectively.

Level	Masses [tons]	Stiffness [MN/m]
Isolation	335	50
1	305	1250
2	285	1125
3	265	1125

Mode	Period [sec]	Frequency [rad/sec]	Effective mass [%]
1	0.98	6.37	99.948
2	0.12	48.38	0.048
3	0.07	89.69	0.002
4	0.05	115.68	<0.001

Table 7.5: Structural characteristics: linearly isolated structure.

Level	Masses	Stiffness Initial - Post	
Isol.	335	250	25
1	305	1250	
2	285	1125	
3	265	1125	

Mode a/a	Period Initial - Post		Frequency Initial - Post		Effective mass [%] Initial - Post	
1	0.47	1.38	13.38	4.55	98.893	99.987
2	0.12	0.13	51.00	48.06	1.036	0.012
3	0.07	0.07	90.54	89.59	0.064	0.001
4	0.05	0.05	115.87	115.66	0.007	<0.001

Table 7.6: Structural characteristics: bilinearly isolated structure (Fy/W=0.05).

Level	Masses	Stiffness Initial - Post	
Isol.	335	150	25
1	305	1250	
2	285	1125	
3	265	1125	

Mode a/a	Period Initial - Post		Frequency Initial - Post		Effective mass [%] Initial - Post	
1	0.58	1.38	10.69	4.55	99.572	99.987
2	0.13	0.13	49.70	48.06	0.404	0.012
3	0.07	0.07	90.10	89.59	0.022	0.001
4	0.05	0.05	115.77	115.66	0.002	<0.001

Table 7.7: Structural characteristics: structure with hybrid isolation (Fy/W=0.03).

The same structure is analyzed for these cases under an earthquake excitation. The time history analysis is performed using the El-Centro earthquake scaled for two levels of peak ground acceleration, pga equal to 0.50 and 0.10 g, respectively, in order to consider the effectiveness of each system under both a severe and a small to moderate earthquake. The damping for the linear system is assumed to be 15% for the first mode and 2% for higher modes, while for the bilinear system, the damping matrix is constructed assuming Rayleigh damping with 2% damping ratios. The damping matrix for the hybrid system is constructed for damping ratios 10% and 2% for the lowest and the highest frequencies, respectively. The yield force of the bilinear system is 5% of the total weight of the structure while that of the hybrid system is 3%. Accordingly, the prior-to-yielding stiffness of the bilinear system is ten times the post-yield stiffness, while that of the hybrid system is six times the post-yield stiffness.

The maximum values for the relative displacement at the isolation level, inter-story deflection, floor acceleration, and base shear force for the three isolation systems and for both pga=0.10g and pga=0.50g are given in Table 7.8 and Table 7.9, respectively.

Isolation type	Base displacement [cm]	Interstory deflections [cm]	Floor accelerations [m/sec^2]	Base shear force [MN]
Linear	1.87	0.056	0.87	0.94
Bilinear	1.27	0.067	1.46	0.84
Hybrid	1.12	0.041	0.85	0.57

Table 7.8: Maximum response values for El-Centro scaled to pga=0.10g.

Isolation type	Base displacement [cm]	Interstory deflections [cm]	Floor accelerations [m/sec^2]	Base shear force [MN]
Linear	9.37	0.278	4.33	4.69
Bilinear	10.0	0.206	3.66	3.03
Hybrid	8.80	0.163	2.79	2.49

Table 7.9: Maximum response values for El-Centro scaled to pga=0.50g.

For comparison purposes, the corresponding maximum response values of the fixed-supported structure with the structural characteristics provided in Table 7.1 and with damping ratios, ξ, equal to 5% are provided below for the two differently scaled levels of the El-Centro earthquake.

El-Centro scaled pga	Interstory deflections [cm]	Floor accelerations [m/sec^2]	Base shear force [MN]
0.10 g	0.11	2.4	1.41
0.50 g	0.57	12.1	7.07

Table 7.10: Maximum response values for El-Centro scaled to pga=0.50g.

The story shear force-interstory deflections for all isolation systems for both levels of earthquake excitation are shown in Figure 7.7 and Figure 7.8.

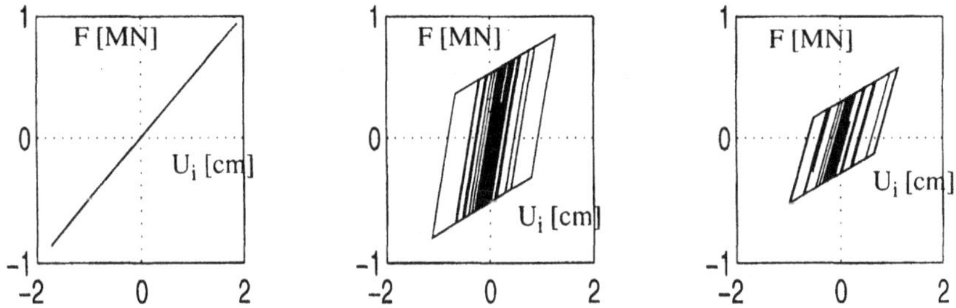

(a) Linear isolation (b) Bilinear isolation (c) Hybrid isolation

Figure 7.7: Base shear force - Relative displacement diagrams for pga=0.10g.

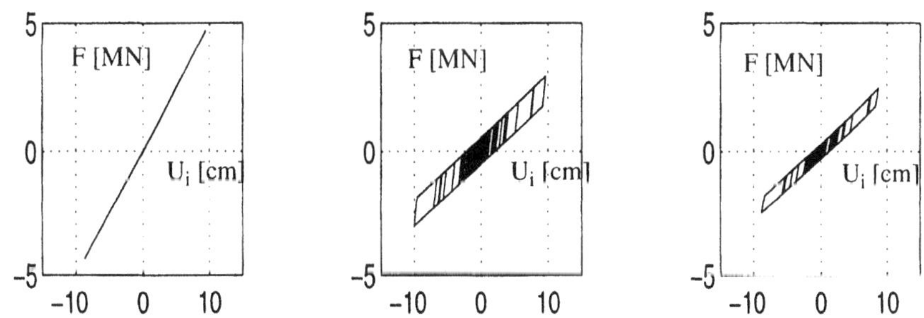

(a) Linear isolation (b) Bilinear isolation (c) Hybrid isolation

Figure 7.8: Base shear force - Relative displacement diagrams for pga=0.50g.

The higher mode effects may be reduced by designing the isolation system to have its initial frequency away from the higher frequencies of the isolated structure in order to avoid resonance. The fundamental frequency of the superstructure with fixed supports may be used as an approximate upper bound from which the initial stiffness of the isolated structure must differ significantly. Initial rigidity is necessary to prevent displacements under other lateral loads, e.g. wind loads, and thus to avoid discomfort to the structure's inhabitants. An alternative to the provision of high initial stiffness is to incorporate an additional energy dissipation mechanism in order to suppress higher mode effects. The hybrid isolation system (LHDRB) provides relatively reduced initial rigidity and an energy dissipation mechanism for minor lateral loads. The LHDRB system reduces higher mode acceleration responses which occur when lead rubber bearings are used. In contrast to lead rubber bearings which dissipate energy only once the isolation system yield force has been exceeded, the system provides a continuous energy dissipation mechanism.

7.6 Linearization of Bilinear Behavior

The bilinear behavior of some highly nonlinear isolation systems, e.g. lead rubber bearings (LRBs), may be linearized for preliminary design purposes. Figure 7.9 shows the typical force-displacement relation of a bilinear system. Nonlinearities due to the hysteretic dissipation of energy may be linearized through the following damping linearization procedure. This is based on the determination of equivalent viscous damping to represent the hysteretic energy dissipation.

7.6.1 Effective stiffness

The initial elastic stiffness provides the necessary initial rigidity for minor lateral loads up to the level of design yield force F_y of the lead plugs. The initial and the post-yield stiffness may be computed considering shear deformations and assuming that there are no bending deformations and effects of axial forces:

$$K_{el} = \frac{G_l\ A_l}{h_{tot}} + \frac{G_r\ A_r}{h_r} \quad , \qquad K_{pl} = \frac{G_r\ A_r}{h_r} \tag{7.7}$$

where G_l and G_r, are the lead and rubber shear moduli, A_l and A_r, the lead and rubber areas, and h_l and h_r, the lead and (total) rubber heights, respectively.

An effective stiffness K_{eff} and an effective damping coefficient ξ_{eff} are very often used to describe the bilinear behavior of the isolation system, and allow an equivalent linear elastic analysis which is useful in the preliminary design stage at least. However, both effective parameters depend on the earthquake excitation and its severity; their determination, therefore, involves several uncertainties.

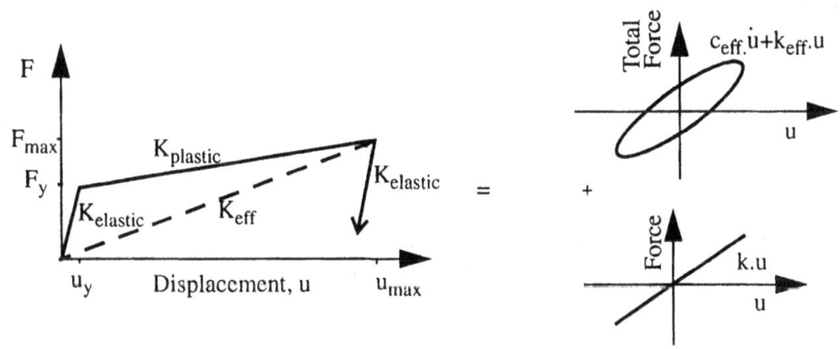

Figure 7.9: Force-displacement bilinear relation and its linearization.

The effective stiffness is defined as the secant modulus joining the extreme positive and negative displacements.

$$K_{eff} = \frac{F_y + K_p\,(u_{max} - u_y)}{u_{max}} = \frac{F_y\,(1 - K_p/K_e) + K_p\,u_{max}}{u_{max}} \tag{7.8}$$

Subsequently, the effective period is given by the expression

$$T_{eff} = 2\,\pi\,\sqrt{\frac{M}{K_{eff}}} = 2\,\pi\,\sqrt{\frac{M\,u_{max}}{F_y\,(1 - K_p/K_e) + K_p\,u_{max}}} \tag{7.9}$$

7.6.2 Effective damping

Effective viscous damping may be defined on an energy equivalence base so as to represent the actual hysteretically dissipated energy. The equivalent damping coefficient ξ_{eff} may be obtained by equating the hysteretically dissipated energy over a complete cycle to the viscously dissipated energy. However, energy dissipation through viscous damping is proportional to velocity, while hysteretic energy dissipation is related to the offsets of reversed cycles and is equal to the kinetic energy converted to thermal energy and dissipated.

A single DOF system with viscous damping C_{eff} and stiffness K_{eff}, shown in Figure 7.10(a), is used to determine an equivalent linear elastic system which in turn represents the hysteretic energy dissipation.

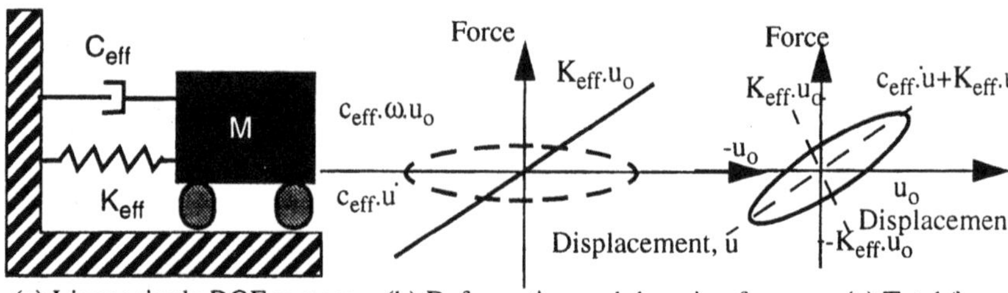

Figure 7.10: Viscously damped linear elastic single DOF system.

Assuming that the displacements for a free vibration of the mass M are of the form $u(t) = u_o \sin(\omega t)$, the velocity is given by: $\dot{u}(t) = u_o \omega \cos(\omega t)$, where ω is the frequency of oscillation. The total force (damping and elastic), is shown in Figure 7.10(c); it is equal to

$$P_{return} = C_{eff} \omega u_o \cos(\omega t) + K_{eff} u_o \sin(\omega t) \tag{7.10}$$

The loss factor h is defined as the ratio

$$h = \frac{(Area\ of\ hysteresis\ loop)}{4\pi\ (Elastic\ energy\ area)} = \frac{E_h}{4\pi E_e} \tag{7.11}$$

and may be determined experimentally by the above-mentioned oscillations.

The hysteretically dissipated energy per cycle is $E_d = \int_0^T P \frac{du}{dt} dt$.

Substituting the expressions for the displacement and velocity, we obtain

$$E_d = \int_0^{\frac{2p}{\omega}} (C_{eff} \omega u_o \cos(\omega t) + \tag{7.12}$$

$$K_{eff} u_o \sin(\omega t))\ \omega u_o \cos(\omega t)\ dt$$

Integration of Equation (7.12) gives the area of the hysteresis loop per cycle $E_d = p\, C_{eff}\, \omega\, u_o^2$. The maximum elastic potential energy is equal to $E_e = 0.50\, K_{eff}\, u_o^2$. Substituting in Equation (7.11), and considering that the natural frequency of the system is equal to $\omega_N = \sqrt{K_{eff}/M}$, the loss ratio h may be related to the damping ratio $\xi_{eff} = \dfrac{C_{eff}}{C_{cr}} = \dfrac{C_{eff}}{2\, \omega_N\, M}$, as follows:

$$h = \frac{\pi\, C_{eff}\, \omega\, u_o^2}{2\, \pi\, K_{eff}\, u_o^2} = \frac{C_{eff}\, \omega}{2\, K_{eff}} = \frac{2\, \xi_{eff}\, \omega_N\, M\, \omega}{2\, \omega_N^2\, M} = \xi_{eff} \left(\frac{\omega}{\omega_N} \right) \qquad (7.13)$$

The frequency of oscillation for the free response is approximately equal to the natural frequency of the system, i.e. $\omega = \omega_N$, and thus the damping ratio ξ_{eff} is, in turn, equal to the loss factor h. The latter is computed experimentally by free vibration tests and is of the order of 10-15% for high damping rubber.

The above 'linearization' of damping is based on several assumptions which may be incorrect in practice. The substitution of hysteretic with equivalent viscous damping, for highly nonlinear systems, involves several uncertainties. First, the hysteretic damping depends on the displacement magnitude since it is equal to the area of cycles, and thus on the input excitation, whereas viscous damping depends on velocity. In addition, the linearization is based on periodic complete cycles of oscillation while the expected irregular response due to an earthquake excitation contains many short cycles with little hysteretic energy dissipation. Small incomplete-cycle viscous damping may overestimate the actual energy dissipation. The experimental results are for free vibration which is not what an isolated structure undergoes. However, the use of an equivalent viscous damping based on an approximate energy equivalence allows a rough approximation of the isolated structure response for preliminary design purposes.

7.6.3 Study of the effectiveness of bilinear behavior linearization

We want to investigate the possibility of using an equivalent viscously damped linear single DOF system to represent the actual nonlinear behavior of a seismically isolated structure with lead rubber bearings. The results indicate that the response is affected by several factors such as the characteristics and intensity of the earthquake, as well as by the system characteristics. This shows the necessity of nonlinear dynamic analysis of the isolated structure at the final design and analysis stages.

A single-DOF system with bilinear behavior is used in the study, assuming it is subjected to six different earthquake excitations (Figure 2.2), scaled for different levels of peak ground acceleration (pga), ranging from 0.10 to 0.70 g.

The system has the following characteristics:

- Mass, M =1672.0 tons
- Elastic stiffness, K_{el}=165.0 MN/m
- Post-yield stiffness, K_{pl}=16.5 MN/m.

Using the Newmark direct-integration method, described in Appendix B, the response of several single-DOF systems with the above characteristics, but with different F_y/W ratios, are calculated. The maximum relative displacements for these systems for all six excitations are shown in Figure 7.11(a).

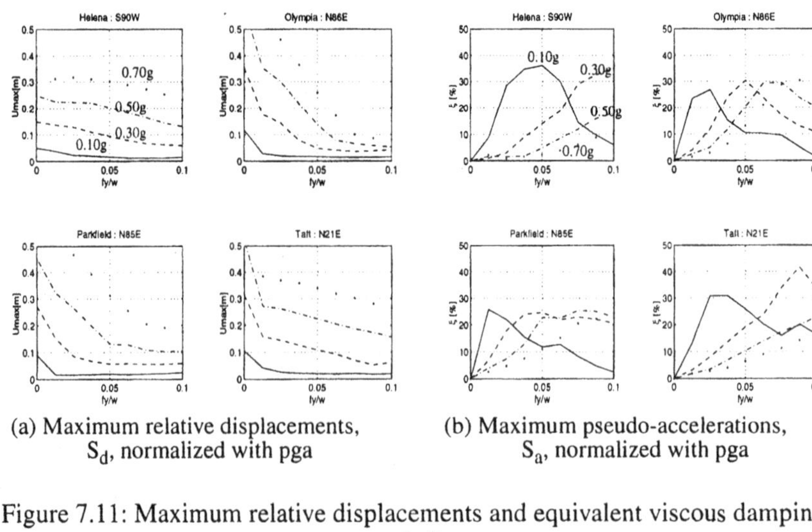

(a) Maximum relative displacements, S_d, normalized with pga

(b) Maximum pseudo-accelerations, S_a, normalized with pga

Figure 7.11: Maximum relative displacements and equivalent viscous damping ratio ξ.

The equivalent viscous damping ratio ξ_{eq}, required to ensure the same maximum relative displacements U_{max} as those of an equivalent viscously damped single-DOF system with the same mass and equivalent effective stiffness K_{eff}, varies with the magnitude of the seismic excitation and the ratio F_y/W. This variation for six different earthquakes is shown in Figure 7.11(b).

We notice from the results obtained from several numerical simulations that the required damping ratio ξ_{eq} of an equivalent linear system varies considerably. In general, the required viscous damping, for an equivalent linear system, is higher under weaker earthquake excitations than under stronger ones, in the range of very low yielding force levels due to the dominance of the prior-to-yielding high initial stiffness of the isolation system. Systems with high levels of yield force are more effective, considering the displacements under stronger earthquake excitations. This is indicated by the higher viscous damping required by an equivalent linear system in order to have the same maximum relative displacement. However, systems with a high level of yield force do not have an energy dissipation mechanism for lower loads, since the hysteretic damping is activated once the isolation system has yielded. In addition, the higher modes may be excited when the degree of non-linearity is also high. Finally, for an intermediate level of yield force, the required viscous damping and the overall effectiveness of the isolation varies considerably according to both structural and earthquake characteristics. In Figure 7.11 we note the reduction of the effectiveness of seismic isolation for very low F_y/W ratios, due to the limited dissipation of energy provided. For a specific earthquake excitation, an optimum value of the yield force F_y can be computed to minimize a particular response quantity. When the quantity of interest is the displacement, the optimum point increases with the severity of the excitation (Figure 7.11(b)). Given the uncertainties and the variations of the characteristics and severity of the expected earthquakes, it is impossible to determine a single optimum damping point.

For a typical lead rubber bearing isolation system, e.g. one with Fy/W=0.05 and K_{el}/K_{pl}=10, the damping ratio is approximately 10-30%. The results show that an equivalent linear elastic, viscously damped system cannot accurately model the actual bilinear behavior of an isolated structure with LRBs. The maximum relative displacements have been computed for a wide range of single-DOF systems with bilinear behavior, assuming for F_y/W the two common cases of 0.02

and 0.05. The previously presented earthquake excitations are used, scaled by the pseudovelocity S_V, given by the ATC3-06 design spectrum for peak ground acceleration (pga), ranging from 0.30 g to 0.60 g. The ATC3-06 is an elastic design spectrum generated in 1984 by Applied Technology Council [6], using records which were scaled to have pga equal to 0.40 g.

The results for the normalized maximum relative displacements U_{max} are compared in Figure 7.12(a) with the corresponding values of (S_d/pga) used for a conventionally designed earthquake-resistant structure, for ξ equal to 5% and 10%, respectively. Also, the resulting maximum accelerations of the bilinear model are compared to the corresponding values of a viscously damped system with ξ equal to 5% and 10%.

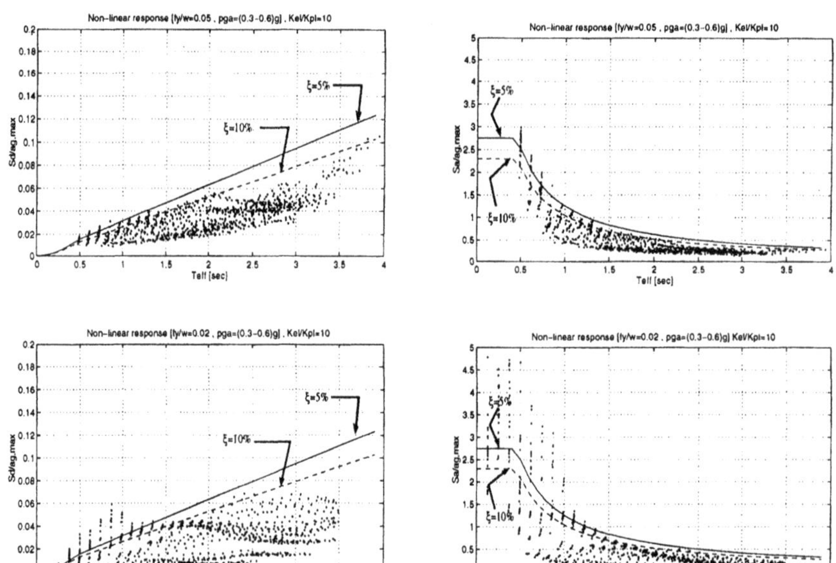

(a) Maximum relative displacements U_{max} normalized with pga

(b) Maximum pseudo-accelerations S_a normalized with pga

Figure 7.12: Comparison of nonlinear response with ATC3-06 response spectra.

We conclude that an elastic design spectrum may be used at the preliminary design stage to take into account the constraint imposed by the maximum relative displacement accommodated for a specific structure. In addition, the results of several linear and non-linear studies indicate that the variation between the linear and non-linear analysis is in many cases less pronounced than the variations among non-linear analyses for different earthquake excitations. Considering the uncertainties associated with the expected earthquake excitations and their specific characteristics, as well as the difficulty involved in a non-linear analysis, we may conclude that the results of a linear analysis are in many cases acceptable. However, higher mode effects and the resulting increased accelerations of the masses cannot be observed using an equivalent linear-elastic model. A nonlinear analysis is required to investigate such effects, in particular for structures with acceleration requirements.

7.7 Effect of the Degree of Non-Linearity

The degree of non-linearity of bilinear systems may be defined as the ratio of the initial stiffness K_{el} to the post-yield stiffness K_{pl}. A high degree of non-linearity may induce higher mode effects and result in large floor accelerations. The degree of non-linearity is also related to the yield force F_y and consequently to the *ratio* F_y/W. Its influence on the system's response is studied using the four-mass system discussed in Section 7.5, the performance of which is compared to that of a typical lead rubber bearing system with a K_{el}/K_{pl} ratio equal to 10.0 and a yield force equal to 5% of the weight of the building. Both F_y/W and K_{el}/K_{pl} vary linearly between 1-10%, and 2-20, respectively. Three different earthquake excitations are used in the analysis, scaled for 0.3, 0.5 and 0.7 g. The structural characteristics of the superstructure are given in Section 7.5. The post-yield stiffness of the isolation system is 25 MN/m. The maximum relative displacements at the isolation level for various degrees of non-linearities, normalized with the corresponding maximum relative displacement for the investigated system with $K_{el}/K_{pl}=10$ and $F_y/W=0.05$ are shown in Figure 7.13. The linear variation of the ratios K_{el}/K_{pl} and F_y/W is plotted on the horizontal axis. We note that they decrease as the non-linearity of the isolation system increases, keeping the post-yield stiffness of the isolation system constant, due to the dominance of the prior-to-yielding high stiffness K_{el}. As the severity of the earthquake decreases, the influence of the non-linearity on the reduction of the maximum relative displacement is more pronounced. This is due to the fact that the effective stiffness of the isolation system is more influenced by the high initial stiffness than by the post-yield stiffness.

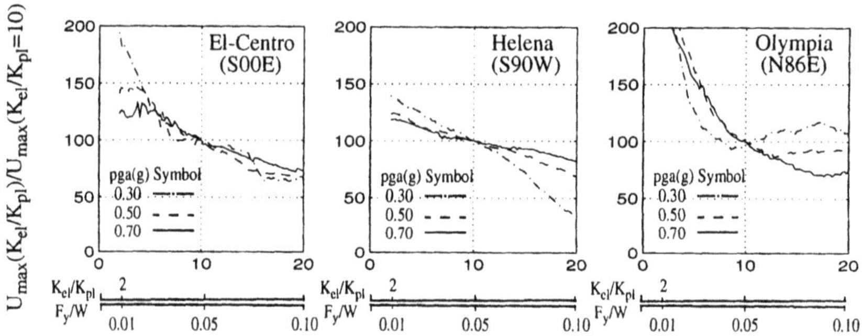

Figure 7.13: Normalized maximum relative displacement at isolation level.

Floor accelerations increase with the increase in the degree of non-linearity, due to the higher effective stiffness of corresponding non-linear systems and due to the increased effective stiffness and higher mode effects excited by the sudden stiffness changes. This is illustrated in Figure 7.14, where the maximum normalized floor accelerations are plotted:

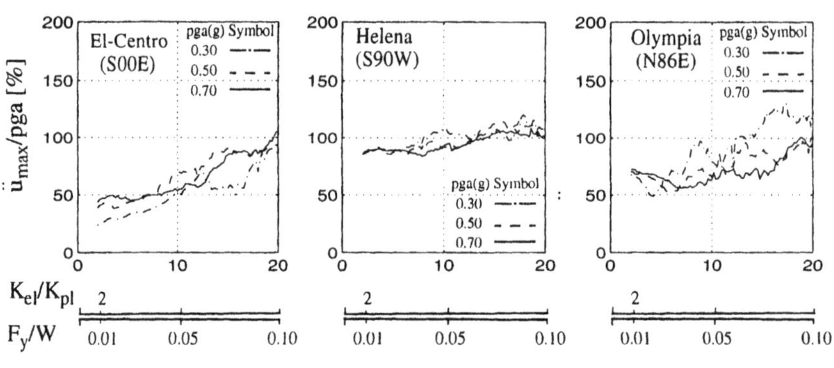

Figure 7.14: Maximum floor accelerations normalized with pga.

A compromise must be made to keep both the maximum relative displacement at the isolation level and the maximum floor accelerations within acceptable limits. The reduction of the former is associated with the maximum displacement to be accommodated by the gap around the structure and the connections of external services; the latter must be controlled to avoid damage to the contents of the structure. The specific project characteristics and design requirements determine which of the two dominates the design.

As previously mentioned, a hybrid model may both reduce floor accelerations and suppress the relative displacement at the isolation level, through additional viscous damping. This is illustrated in Figure 7.15 where the viscous damping of the system is increased linearly from *2%* at $K_{el}/K_{pl}=10$, to *10%* at $K_{el}/K_{pl}=6$. Additional viscous damping may be provided either by the inherent damping properties of high-damping rubber of an elastomeric bearing, or via the incorporation of an actual viscous damper.

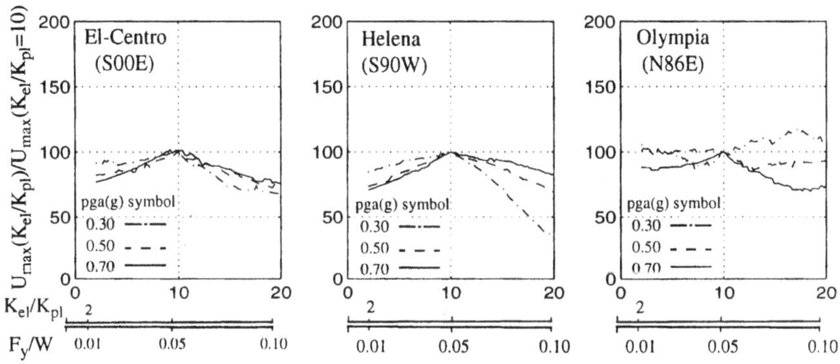

Figure 7.15: Normalized maximum relative displacements (additional viscous damping).

The increase of the relative displacements at the isolation level with the decrease of the degree of non-linearity may be suppressed, as illustrated in Figure 7.15. Reductions in floor accelerations may also be achieved by decreasing the non-linearity of the isolation system, as shown in Figure 7.16.

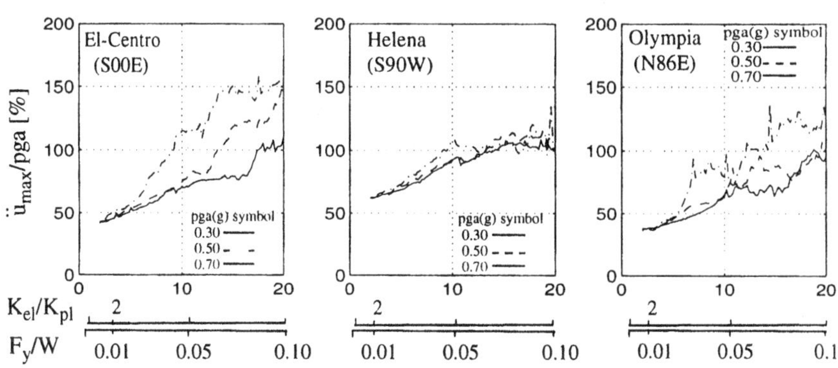

Figure 7.16: Maximum floor accelerations normalized with pga (additional viscous damping).

It is evident from the above discussion that when the reduction of floor accelerations is an important design parameter, the degree of non-linearity must be reduced as well as F_y/W. The latter is defined by the minor lateral loads for a specific project. Thus, using a particular F_y/W ratio, a proper K_{el}/K_{pl} ratio must be selected. When building contents protection dominates the design requirements, a lower degree of non-linearity must be used, combined with additional viscous damping to suppress the large relative displacement at the isolation level. The use of highly nonlinear isolation systems may result in increased higher-mode effects which may significantly increase mass accelerations and consequently the inertia forces induced in the structure and its contents. The contribution of higher modes is enhanced by sudden changes of the isolation system stiffness during reversed loads. Higher-mode effects may be reduced by designing the isolation system to have its initial, i.e. prior-to-yielding, frequency away from the higher frequencies, to avoid resonance. These, together with the resulting increased mass accelerations, cannot be observed using an equivalent linear-elastic model.

7.8 Effect of Superstructure Stiffness

The stiffness of the superstructure and in particular its relative difference from the stiffness of the isolation system significantly affects the efficiency of seismic isolation. The degree of isolation, defined as the ratio of the period of the isolated structure to the period of the superstructure when fixed supported, is directly related to the ratio of the corresponding stiffness for both types of boundary conditions. In general, as the stiffness of the superstructure increases, the large relative displacement at the isolation level slightly increases, due to the increase of the fundamental period. The inter-story displacements are substantially reduced due to the increased stiffness of the superstructure; floor accelerations are reduced due to the increase of the fundamental period. Finally, the base shear forces increase slightly due to the increase in the relative displacement at the isolation level.

A 3-DOF system mounted on a linear isolation system, with characteristics as given in Figure 7.17, is analyzed as if during the El-Centro earthquake, to illustrate the above effects as the stiffness of the superstructure changes.

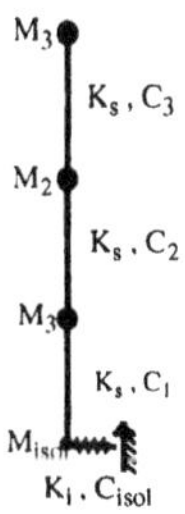

Level	Masses [tons]	Stiffness [MN/m]	Damping ratio ξ [%]
Isolation	335	50	15
1	305	varying from 500 to 5000	2
2	285	varying from 500 to 5000	2
3	265	varying from 500 to 5000	2

Figure 7.17: Structural characteristics of the isolated structure.

The ratio of the superstructure stiffness K_s to the isolation system stiffness K_i varies from 10 to 100. The maximum values of the quantities of interest are given in Figure 7.18 and Figure 7.19. In particular, the change of the fundamental period of the isolated structure as a function of K_s/K_i, normalized by the value of the fundamental period for K_s/K_i=10, is shown in Figure 7.18(a). The fundamental period of the multi-DOF system approaches the value of the period of a single-DOF system when the superstructure is assumed to be infinitely rigid, and is indicated by a dashed line. The maximum values of relative displacements at the isolation level and the inter-story deflections, normalized with the corresponding value at $K_s = 10\ K_i$, are also shown. The relative displacement at the isolation level increases by less than 5% when the superstructure stiffness becomes ten times its initial value, while the inter-story displacements decrease substantially by 90%.

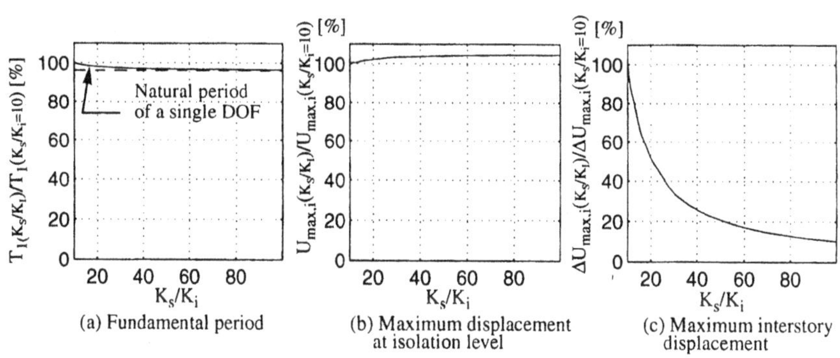

Figure 7.18: Effect of superstructure stiffness on displacements.

The maximum floor accelerations and base shear force, with varying degrees of isolation, are shown in Figure 7.19. They decrease by approximately 10%, while the base shear forces increase by less than 5%.

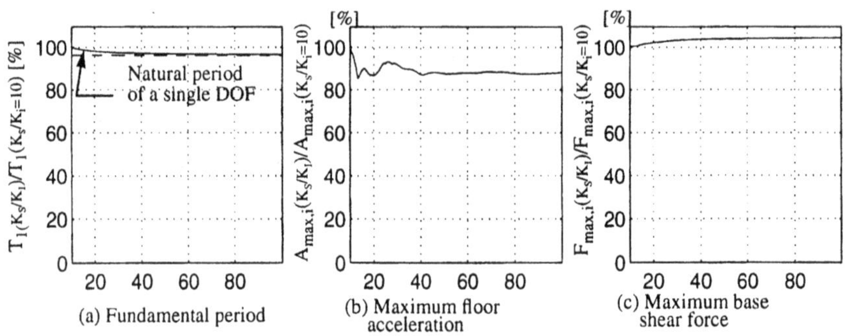

Figure 7.19: Superstructure stiffness effect on floor accelerations and base shear forces.

It is evident that the superstructure must be as stiff as possible to ensure a better performance when seismic isolation is used. The larger the difference between the fundamental frequency of the seismically isolated structure and its fundamental period when fixed supported, the better the response of the isolated structure.

7.9 Modeling of Seismically Isolated Bridge Structures

Typically, only the superstructure of a bridge structure is seismically isolated. Therefore, the bridge structure can be simulated using the following simple model (Figure 7.20(a)). For each pier the mass of the pier m_p and the part of the superstructure mass corresponding to that pier m_b can be lumped into a two-DOF system.

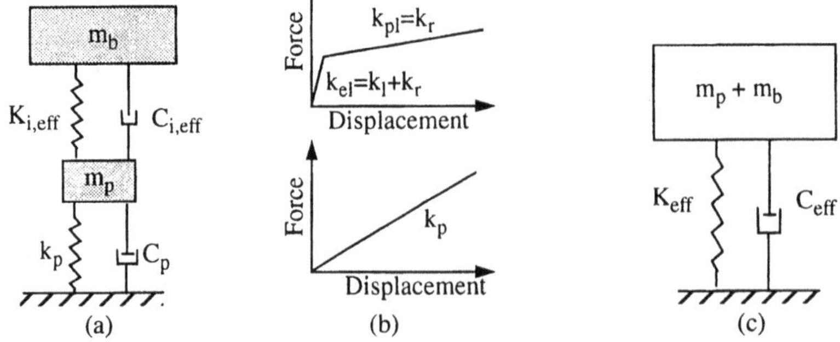

Figure 7.20: Simplified model of pier with bilinear isolation.

The effective stiffness of the simplified model can be computed from the stiffness of the pier k_p and the effective stiffness of the isolation system $K_{i,eff}$ at that pier. If the isolation system is bilinear, such as when using lead-rubber bearings, which is shown in Figure 7.20(b), the effective stiffness of each isolator can be computed by the following equation:

$$k_{i,eff} = \frac{F_{max}}{U_{max}} = \frac{F_y\left(1 - \frac{K_{pl}}{K_{el}}\right) + K_{pl}\,U_{max}}{U_{max}} \qquad (7.14)$$

Often the isolators at a certain pier have the same geometric and mechanical characteristics; in that case the effective stiffness of the isolation system can be computed by simply adding the effective stiffness of each isolator.

$$K_{i,eff} = \sum k_{i,eff} \qquad (7.15)$$

Then the effective stiffness of a pier, which is modeled by a single DOF (Figure 7.20(c)), can be computed as

$$K_{eff} = K_{i,eff}\frac{1}{\left(1 + \frac{K_{i,eff}}{k_p}\right)} \qquad (7.16)$$

The longitudinal effective overall stiffness can be approximated by summing the effective stiffnesses of all piers K_{eff}. The effective mass can be computed by summing the mass of each pier and the bridge superstructure. Therefore, an effective period T_{eff} in the longitudinal direction can be estimated and used in a preliminary design.

$$T_{eff} = 2\pi\sqrt{\frac{\sum m_p + m_b}{\sum K_{eff}}} \qquad (7.17)$$

However, the above equations can be used only to give very rough approximations of the stiffness and period of a seismically isolated bridge. For the final design and analysis, a detailed bridge model should be used to perform time history dynamic analyses during the design earthquakes.

The above simple modeling is applicable for the case in which only the superstructure of a bridge has been seismically isolated, which is the most common practise due to its cost effectiveness and ease of implementation. For bridges where overall seismic isolation is used, more sophisticated models must be employed.

7.10 Ductility of Seismically Isolated Structures

The seismic loads considered in conventional earthquake resistance design are substantially reduced using the reduction factors provided in the codes. The reduction of the seismic forces is due mainly to the inelastic deformations which dissipate energy. The superstructure of an isolated structure may also have some inelastic deformations under a very strong excitation. In addition, the softening of the superstructure from the inelastic deformations lengthens the period of the superstructure. In contrast, in the case of a seismically isolated structure in a very strong earthquake, although energy is dissipated through inelastic deformations when the superstructure begins to yield, the softening of the superstructure has negative effects on the efficiency of seismic isolation. The ductility capacity of a seismically isolated structure is much lower than that of a fixed supported one. This difference is reflected in the different reduction factors adopted for conventionally fixed supported and seismically isolated structures. Appropriate factors for the latter are under investigation and no agreement on their values has been reached. The writer's opinion is that the reduction factors may be used only for the maximum credible earthquake; since no damage is accepted for the design earthquake, no reduction factors should be used.

Chapter 8

A Preliminary Design Methodology

8.1 Introduction

A preliminary design methodology for bilinear elastomeric isolation systems, those with lead plugs in particular, is presented here. The purpose of the methodology is to provide only a preliminary guidance for the design of an isolated structure, which involves several parameters which are dependent on each other. A trial and error procedure with adjustments and corrections may be required finally to obtain an adequate design of the isolation system. Although it is a creative procedure, the design of an isolated structure is complicated and requires significant experience and extensive engineering judgement. The dominant design parameters of the proposed methodology are the level of yield force F_y of the isolation system, the maximum allowable relative displacement at the isolation level U_{max}, the location and dimensions of the isolators, the non-linearity ratio K_{el}/K_{pl} and the distribution of seismic forces at the isolators.

The level of the yield force F_y of the isolation system is determined by the minor lateral loads to be sustained prior to yielding, and is usually expressed as a percentage of the weight of the structure. Practical restrictions constrain the level of maximum allowable relative displacement at the isolation level U_{max}, and consequently the degree of isolation. The ratio K_{el}/K_{pl} is an important factor for the reduction of floor accelerations and must be selected according to the design requirements. As discussed in Chapter 7, this ratio must be relatively low when protection of building contents dominates the design. Finally, proper selection of locations and dimensions of the isolators, considering their axial forces under service loads, can suppress torsional and rocking effects. Torsional effects result from eccentricities between the centers of stiffness and mass of the building. These eccentricities are due to the asymmetric configuration of the structural members, especially the very stiff elements such as the elevator core and the stairwells, and the asymmetric horizontal distribution of gravity loads. Although the proposed preliminary design methodology involves isolation systems with bilinear behavior applied to relatively stiff structures, the primary issues of concern are directly pertinent to most isolation systems.

8.2 A Feasibility Study for the Use of Seismic Isolation

At the preliminary design stage, a feasibility study must be carried out to investigate the plausibility and effectiveness of seismic isolation for the structure under consideration. When the protection of a building's contents and the avoidance of loss of functionality are the dominant design objectives for a building located in a high seismicity area, and thus suitable for seismic isolation, it is almost always the most effective and, in the long term, most economic design method. However, it is not suitable for all types of structures since its effectiveness is closely related to the superstructure characteristics.

As previously mentioned, seismic isolation is suitable for stiff low- to medium-rise buildings, bridges, and some special relatively stiff structures, such as nuclear reactors; it is not usually appropriate for flexible high-rise buildings. A desirable characteristic of a structure to be seismically isolated is a low aspect ratio, i.e. it is relatively squat. Moreover, the seismicity of the area indicates the significance of seismic loads and the necessity for the incorporation of an appropriate system to account for these loads. Other minor loads such as wind loads must also be considered in determining the applicability of seismic isolation. Wind and other non-earthquake lateral loads, typically, should not exceed 10% of the weight of the structure. In addition, site and foundation soil characteristics reveal the expected dominant frequencies of earthquake excitations and indicate the suitability and effectiveness of the method. In general, the stiffer the soil, the more effective the seismic isolation of a structure founded on such soil. Finally, even when the structure of concern is suitable for seismic isolation, the proximity of surrounding structures to it may prevent isolation. The site and the surrounding structures should permit the anticipated maximum horizontal displacements at the isolation level, which are usually of the order of 10-30 centimeters. As previously mentioned, a seismic gap at the isolation level must be ensured; in some cases it may not be available if adjacent structures are too close to the structure concerned.

For seismic upgrading of existing structures for which seismic isolation is to be used, the difficulty and extent of modifications involved in its incorporation must also be considered in the feasibility study. The location of the isolation level may be restricted due to the structure's characteristics. In many cases of retrofitting of existing structures, additional measures may be necessary to upgrade the strength and ductility of the existing foundations and superstructure.

8.3 Determination of Peak Ground Acceleration

The level of seismic loading indicates its importance and must, therefore, be determined. The seismic input generally adopted consists of two levels of seismic excitation: a design earthquake, which is likely to occur once or twice during the useful life of the structure, and a maximum credible earthquake, which is the most severe earthquake expected in a particular region given the a priori knowledge of the local geological history. According to the provisions of most codes for seismically isolated structures, these two levels of seismic excitation must be considered in the design. The Uniform Building Code requires the isolated structure of being designed and analyzed using a design earthquake loading with a 10% probability of being exceeded in 50 years, and a maximum credible earthquake with a 10% probability to be exceeded in 250 years. The isolated structure must respond elastically, i.e. without damage, for the design earthquake. For the maximum credible earthquake, limited inelastic deformation is allowed in the superstructure. The earthquake loadings must be determined considering the seismicity of the region, the importance of the structure and site conditions.

In a preliminary design procedure, based on the maximum expected displacement at the isolation level, it is adequate to use a linear design spectrum, having first determined the peak ground acceleration (pga) for the maximum credible earthquake. The estimation of pga may be done from acceleration attenuation curves, by locating the active seismic sources of the area and using the magnitude of the maximum credible earthquakes and the corresponding epicentral distances. The acceleration attenuation curves have been derived after extensive seismological research and studies on the soil characteristics of different regions, and provide pga as a function of earthquake magnitude and epicentral distance. Attenuation curves for different regions with similar geological characteristics may instead be used when those for a particular area are not available. Alternatively, the available expected pga given in the codes used in conventional earthquake resistance design, may be used to determine the pga for the design and maximum credible earthquakes of the isolated structure.

8.4 Design of the Superstructure

The superstructure must be initially designed independently of the isolation system, although some of its structural characteristics may influence the efficiency of the system. In Chapter 6 it has been demonstrated that the stiffer the superstructure, the more efficient the seismic isolation system is, since the difference between the fundamental period of the isolated structure and the fixed supported superstructure increases with the inserted flexibility. The deformations of the superstructure are substantially decreased according to this difference, due to the actually rigid body motion of the whole structure. The masses accelerations are also reduced with the inserted flexibility, due to the avoidance of resonance. Thus, the superstructure must be designed to be as stiff as possible within reasonable economic cost and architectural impact.

The superstructure must be analyzed for gravity dead and live loads, used to generate the required loading combinations in the analysis and design stages. Two loading combinations, namely maximum gravity loads and service gravity loads, must be determined by simple superposition of appropriately factored dead and live loads. Service gravity loads must also be combined with seismic, wind and/or any other non-gravity loads to generate the corresponding loading combinations. When special design details of the bearings do not allow the transmission of bending moments from the columns to the bearings, which is typically the case, the degrees of freedom corresponding to the supports' displacements must be restrained. In contrast, the ones corresponding to the three rotations must be released. This modeling is based on the assumption that there is a diaphragm, above the isolation level, which is sufficiently rigid in its plane to prevent relative displacements of the support nodes.

The configuration of the vertical members of the superstructure determines the potential locations of the isolators. Torsional effects are avoided by coinciding the center of stiffness with the center of mass of the superstructure. In practice, such effects cannot be completely eliminated. Therefore, in order to minimize them it is preferable to have the isolators at the outer edges of the structure, as far away as possible from the center of stiffness. The vertical configuration of the isolation system must be determined and a fully stiff diaphragm must be constructed above the isolation level, to distribute lateral loads uniformly to the isolators, taking advantage of the large in-plane stiffness of the diaphragm. The plan dimensions of the bearings may be computed from the corresponding vertical loads which they must sustain. When the construction of a diaphragm is not possible, the bearings must be designed in proportion to the magnitude of the lateral forces carried by the members above them. It is difficult to take into account such design issues due to the uncertainties involved; therefore, the construction of a diaphragm above the isolation level must be anticipated. Note that here, the existence of a rigid diaphragm is assumed. Finally, the selected location of the bearings must be such that access to them is enabled for inspection and possible replacement purposes.

8.5 Selection of a Suitable Isolation System

The type and structural characteristics of the building under consideration, as well as the given design requirements, must be considered in the selection of an appropriate isolation system. Non-linear systems, e.g. those with hysteretic or frictional behavior, are more effective in reducing large relative displacements and inter-story deflections, particularly under extreme seismic loads. However, highly non-linear isolation systems are not appropriate when the protection of the structure's contents is an essential design parameter, since they may produce high mode effects resulting in high floor accelerations. The latter may be reduced when the high initial, i.e. prior-to-yielding, stiffness of the isolation system is significantly lower than the stiffness of the superstructure. This ensures a sufficient degree of isolation for the loading level prior to yielding. This difference in stiffness is easily attained in very stiff superstructures with a high fundamental frequency. Alternatively, linear systems may be used to suppress higher mode vibrations and reduce floor accelerations. In addition, the continuous presence of a dissipation mechanism, even for minor lateral loads, may be essential in some cases where protection must be ensured under microtremors. The designer must also consider the availability and cost of each different type of seismic isolation system as well as the experience of those who provide it. The most commonly used seismic isolation systems have been presented in Chapter 4, where the use of a hybrid isolation system has been proposed, to achieve an optimum compromise between the extreme cases of an approximately linear elastic and a highly non-linear hysteretic system. The hybrid isolation system, which has relatively low initial stiffness, has an increased degree of isolation prior to yielding, compared to a highly nonlinear system with significant abrupt stiffness changes. Finally, the objective and available funds for a particular project also determine which isolation system should be used. For example, cost considerations dominate the selection of an appropriate design in low-cost dwellings, while for very important structures with expensive equipment, such as nuclear plants and computer centers, seismic performance and protection requirements prevail in the design.

8.6 Preliminary Design of the Isolation System

The response of a seismically isolated structure under a strong earthquake excitation is approximately equal to that of the structure with the superstructure above the isolation level treated as rigid. This has been demonstrated in Chapter 6 for the case of an isolation system inserted at the base of a structure. Thus, a single degree-of-freedom system, where the mass is that of the superstructure and the stiffness is that of the isolation system, may represent the structure adequately, at least for preliminary design purposes. Note that the results of a non-linear dynamic analysis may be very different for the same structure and isolation system, under different seismic excitation scaled for the same level of pga. Therefore, considering the uncertainties associated with both the seismic loading and the response of a seismically isolated structure, it is sufficient to use an equivalent linear elastic viscously damped model and a design response spectrum in the preliminary design stage.

The design of the isolation system must be based on a compromise between the degree of the isolation to be provided to the structure and the maximum allowable relative displacement at the isolation level. The more isolated a structure, i.e. the larger the flexibility inserted at the isolation level, the lower the seismic forces, inter-story deflections and floor accelerations. On the other hand, the relative displacements are larger and thus practical constraints are imposed on the degree of isolation. In addition, the yield level is constrained by minor lateral loads, such as wind loads, to avoid discomfort to the occupants whenever the structure is subjected to such loads. When significant constraints are imposed on the acceptable maximum relative displacement, a high degree of isolation may be reached using auxiliary energy dissipation mechanisms, such as viscous dampers, to suppress the displacement at the isolation level. However, excessive damping may excite higher modes and result in higher floor accelerations, which must be avoided.

8.6.1 Estimation of the maximum allowable displacement

At the preliminary design stage, the maximum relative displacement of a structure may be approximately determined, considering practical restrictions due to the location of the structure and the preservation of facilities such as elevators, pipes etc. Reduction of the maximum allowable relative displacement at the isolation level may result in savings due to the reduced required seismic gap around the structure, and the flexibility of the connections to external services to accommodate such displacement.

The maximum allowable shear strains of the elastomeric bearings may be determined by the manufacturer or by laboratory tests and must be used to check the selected dimensions of the isolators. The allowable shear strain used in the design of bridge bearings is 20% of the failure strain, ε_f in a simple tension test (Skinner et al [103]). However, bridge bearings have to carry well-defined service loads which act continuously. Instead, the loads used for the design of seismically isolated structures are extreme seismic events which are unlikely to happen more than once or twice during the lifetime of the structure. Therefore, for earthquake loadings the allowable shear strains may be higher than the shear strains under frequent loadings. Allowable shear strains are typically of the order of 0.5 for design earthquakes and higher for extremely rare events; tests indicate rupture of the rubber for strains of the order of 500%.

8.6.2 Estimation of yield force F_y of isolators

The yield force of the isolation system must be determined considering minor lateral loads such as wind loads. It is, indeed, usually considered to be equal to these loads. Initial high stiffness is provided by the lead of the LRBs which may be used either under each column or only under selected columns at the perimeter of the building. The latter case is preferable when relatively low prior-to-yielding stiffness is anticipated, since it provides higher resistance to torsion. When LRBs are used under each column, the lead plug dimensions may be easily determined using the axial forces for service gravity loads. The analysis of the superstructure under the service gravity loading combination defines the axial loads at each bearing location which may be used properly to determine the size of the lead plugs to avoid torsional effects. The diameter of each lead plug may be determined considering the minor loads to be carried by the lead plugs and using the lead yield strength, in order to have initial rigidity under these loads. The yield strength of lead $\sigma_{l,y}$ is approximately 10.5 MPa, while the ratio of the yield force F_y, to the total weight of the structure under service loads W_s is typically chosen to be 5%. As shown in Chapter 6, a lower value of F_y/W_s ratio, e.g. 3%, may be more appropriate, particularly for floor acceleration reductions. The required area and diameter of the lead plug may be determined using the computed axial load at the location of each bearing and the ratio of F_y to the total weight of the structure under service loads, W_s.

$$A^i_{lead} = \frac{F_y}{\sigma_{lead,\,y}} \frac{W^i_s}{W_s} \tag{8.1}$$

$$d^i_l = \sqrt{\frac{4\,A^i_l}{\pi}} = \sqrt{\frac{4\,F_y}{\pi\,\sigma_{lead,\,y}} \frac{W^i_s}{W_s}} \tag{8.2}$$

Using the above formula to compute the required lead plug diameter, the center of mass for the service loading combination will automatically coincide with the center of stiffness prior to yielding. Mayes et al. [76] estimate that the lead plug diameter must lie in the range of one-third to one-sixth of the bearing diameter, i.e. $D_r/6 < d_{lead} < D_r/3$, and not more than two-thirds of the total rubber height h_{tot}, values which are not yet known at this stage of the design, and must therefore be checked at a later stage.

8.6.3 Determination of required effective stiffness

Seismic isolation is more effective when the fundamental period of the fixed supported superstructure and that of the isolated structure are well separated, which in turn results in rigid body motion of the superstructure with small floor accelerations and inter-story deflections. The objective of the design is to make seismic isolation more efficient by increasing the inserted flexibility to ensure a large effective fundamental period, under the restriction of maximum acceptable displacement at the isolation level. The preliminary design must primarily address this issue, using a simplified single DOF system. To determine the required characteristics of the isolation system, the effective stiffness and damping may be used at the preliminary design stage to provide rough approximations of the overall response of the isolated structure.

Having estimated the maximum acceptable displacement (U_{max}), the pga for the maximum credible earthquake, the ratio F_y/W, the soil coefficient S and the damping coefficient B are used to compute the required effective period T_{eff} from a normalized design spectrum. The design spectrum, according to the Uniform Building Code (UBC), is defined by the following expressions:

$$C = min\left(\left(\frac{1.25\ S}{T^{2/3}}\right), 2.75\right) \quad , \qquad S_a = \frac{C\ pga}{B} \tag{8.3}$$

where S is the site coefficient for considering the foundation soil characteristics, and B is the damping coefficient. The site coefficient takes values of 1.0, 1.2 and 1.5 for soil types S_1, S_2 and S_3, respectively. Although these values have been adopted by the 'Earthquake Design' of UBC for conventional fixed supported structures, they are more appropriate in the preliminary design than those given in provisions for seismically isolated structures. The latter have been conservatively overestimated for the case of a simple static response procedure, and their use leads to the selection of ineffective seismic isolation systems. The damping coefficient may be determined using an approximate estimation of the minimum equivalent viscous damping ratio ξ. In particular, B may be considered to be 1.0, 1.2, 1.5 and 1.7 for damping ratio ξ equal to 5%, 10%, 20% and 30%, respectively. Considerable research has been done on the expected seismic events, the results of which are currently used in the conventional earthquake resistance design, and which may also be used in the preliminary design of seismically isolated structures. Once T_{eff} has been determined and the ratio (K_{el}/K_{pl}) has been specified, the initial stiffness K_{el} and the post-yield stiffness K_{pl} of the isolation system may subsequently be computed.

When a design spectrum is used, the equivalent viscous damping must also be determined. Although some inelastic deformation of the superstructure is allowed for the maximum credible earthquake, the energy dissipation mechanism must be limited to that absorbed by the seismic isolation system, not allowing ductility reductions for safety purposes. As discussed in Chapter 6, we have tried to investigate whether the bilinear response of a seismically isolated structure with lead rubber bearings can be characterized by an equivalent viscously damped system. A single degree-of-freedom system has been used to model the physical problem; the obtained results showed pronounced variations. However, when we compared the computed maximum displacements with those obtained from the ATC3-06 design spectrum for a 10% damping ratio, we observed that the computed displacements were well enveloped by those obtained from the design spectra. The variation of the equivalent viscous damping for a hybrid system is less than that of an LRB which has a purely hysteretic energy dissipation mechanism. Having estimated a damping ratio ξ_{eff}, the damping coefficient B may be subsequently determined.

A near-fault effect coefficient N is also defined in the current provisions for the static procedure used for seismic isolation, to consider the effect of long components of the ground motion in the design displacements, due to the proximity of the structure to an active fault. This coefficient increases the design displacement by 50% and 20% when the structure is located within 5 or 10 km, respectively, from the active fault. This coefficient is conservatively high, mainly to limit the use of the simple static procedure. In this preliminary design procedure it has been ignored since its use requires high stiffness due to the substantial increase of the estimated maximum displacement. The possibility of dominance of long components of the ground motion must be taken into account during the dynamic analysis of the isolated structure, using excitations with such characteristics. Adoption of the relatively high values of the near-fault coefficient would result in a significant increase in the expected displacement, using the design spectrum, and consequently to less flexible isolation systems. In most cases, lower flexibility of the isolation system not only reduces its effectiveness but may also result in resonance between the isolation system and the ground motion. In extreme cases, e.g. very soft soils, where the dominant ground motion period is so high that it is impossible to reach the necessary difference from the fundamental period of the structure, seismic isolation should not be used.

The pseudo-velocity S_v in the range of the fundamental period of a typical seismically isolated structure is constant. The spectral acceleration is therefore proportional to the inverse of the period T and not to $T^{(2/3)}$. Thus, in the denominator of the expression for the coefficient C in Equation (8.3), T should be used instead of $T^{(2/3)}$. The latter has been conservatively adopted in codes for very flexible structures. In addition, for a typical fundamental period of an isolated structure the coefficient C is less than 2.75. Therefore, the pseudo-velocity S_v may be computed as a function of C, by the expressions

$$C = \frac{1.25\ S}{T} \quad , \qquad S_v = \frac{1.25\ S}{2\ \pi\ B}\ pga \tag{8.4}$$

The maximum relative displacement is then computed as:

$$U_{max} = S_d = \frac{S_v}{\omega_{eff}} = \frac{1.25\ S\ pga\ T_{eff}}{4\ \pi^2\ B} \tag{8.5}$$

Having determined the maximum acceptable relative displacement U_{max}, the ratio of the yield force to the weight of the structure F_y/W_s, pga and equivalent viscous damping, the required effective period T_{eff} may be computed as

$$T_{eff} = \frac{4\ \pi^2\ B\ U_{max}}{1.25\ S\ pga} \tag{8.6}$$

The effective stiffness K_{eff} may be expressed, using the force-displacement relation of a bilinear system, as a function of U_{max}:

$$K_{eff} = \frac{F_{max}}{U_{max}} = \frac{F_y\left(1 - \frac{K_{pl}}{K_{el}}\right) + K_{pl}\ U_{max}}{U_{max}} \tag{8.7}$$

The effective period T_{eff} is then given by

$$T_{eff} = 2\ \pi \cdot \sqrt{\frac{M}{K_{eff}}} = 2\ \pi \sqrt{\frac{M\ U_{max}}{\left(1 - \frac{K_{pl}}{K_{el}}\right) F_y + K_{pl}\ U_{max}}} \tag{8.8}$$

We set Equations (8.6) and (8.8) to be equal and obtain

$$T_{eff} = \frac{4\ \pi^2\ B\ U_{max}}{1.25\ S\ pga} = 2\ \pi \sqrt{\frac{M\ U_{max}}{\left(1 - \frac{K_{pl}}{K_{el}}\right) F_y + K_{pl}\ U_{max}}} \tag{8.9}$$

A K_{el}/K_{pl} ratio must be assumed and then the above equation must be solved for the post-yield stiffness K_{pl}. For lead rubber bearings, the ratio K_{el}/K_{pl} is typically 10, while for lead high-damping rubber bearings it may be lower (close to 6).

$$K_{pl} = \frac{M\ S^2\ pga^2\ -25.266\ B^2\ F_y\ U_{max}\left(1 - \frac{K_{pl}}{K_{el}}\right)}{25.266\ B^2\ U_{max}^2} \tag{8.10}$$

The post-yield stiffness K_{pl} may be specified for each particular isolator using the axial load W_s^i which corresponds to a service load combination, since the isolators act as springs in parallel. It is convenient to express the derived equation as the ratio of the post-yield stiffness K_{pl} to the axial load at a particular isolator:

$$\frac{K_{pl}^i}{W_s^i} = \frac{S^2\ pga^2\ -247.86\ B^2\ (F_y/W_s)\ U_{max}\left(1 - \frac{K_{pl}}{K_{el}}\right)}{247.86\ B^2\ U_{max}^2} \tag{8.11}$$

For a very low effective period, which is almost never the case, the post-yield stiffness of each isolator may be determined by using the following equation, obtained using C=2.75, to determine the maximum displacement:

$$\frac{K_{pl}^i}{W_s^i} = \frac{2.75\ pga\ -9.81\ B\ (F_y/W_s)\left(1 - \frac{K_{pl}}{K_{el}}\right)}{9.81\ B\ U_{max}} \tag{8.12}$$

Once the maximum acceptable relative displacement U_{max} and the yield force F_y have been determined, and assuming a particular non-linearity ratio K_{el}/K_{pl}, the required post-yield stiffness may be calculated from the above formulas. The global parameter K_{pl}^i/W_s^i, may be used to compute the required stiffness of each individual isolator, by multiplying the computed ratio by the corresponding vertical loads. Then, the center of post-yield stiffness K_{pl} will coincide with the center of mass of the superstructure under service loads, thus avoiding torsional effects.

The variation of the required ratio of post-yield stiffness to axial load K_{pl}/W_s^i is shown in Figure 8.1 for two different cases, as a function of the maximum displacement U_{max}. In the first case (Figure 8.1(a)), the assumptions are: a stiff and shallow foundation soil, i.e. S_1 soil coefficient, minor lateral loads equal to 3% of the total service weight of the structure, a ratio of initial to post-yield stiffness equal to 6, approximately 10% equivalent viscous damping, and that the site is located more than 15 km from the active fault concerned. In the second case (Figure 8.1(b)), it is assumed that the structure is founded on a soil profile of soft to medium stiff clay with a limited depth of soft clay, i.e. S_3 soil coefficient, that minor lateral loads are of the order of 5% of the total service weight, that K_{el}/K_{pl} is equal to 10, that equivalent viscous damping is 20% and that the active fault is situated within less than 10 km of the structure. The required values of $(K_{pl}/W_s)^i$ are plotted in Figure 8.1 for each of the above-mentioned cases, for three different peak ground accelerations, 0.3, 0.4 and 0.5 g.

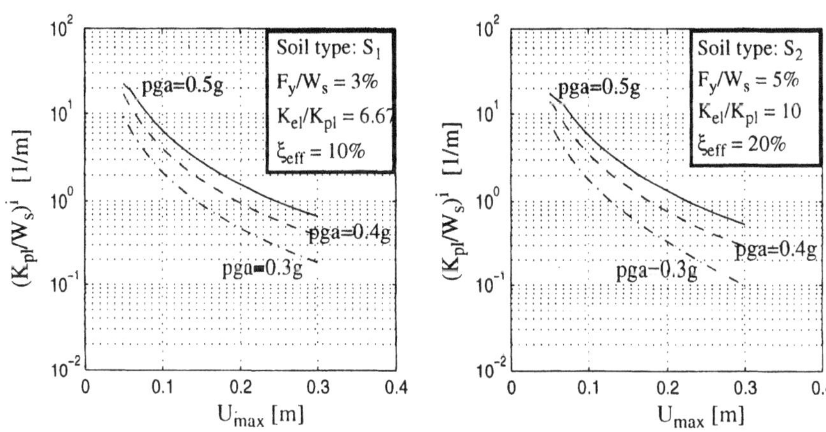

Figure 8.1: Design charts for $(K_{pl}/W_s)^i$, based on U_{max}.

The values of the ratio of post-yield stiffness $(K_{pl}/W_s)^i$ to the maximum allowable displacement U_{max}, as derived by the above charts, give the effective period values T_{eff} using Equation (8.8). The charts are shown in Figure 8.2.

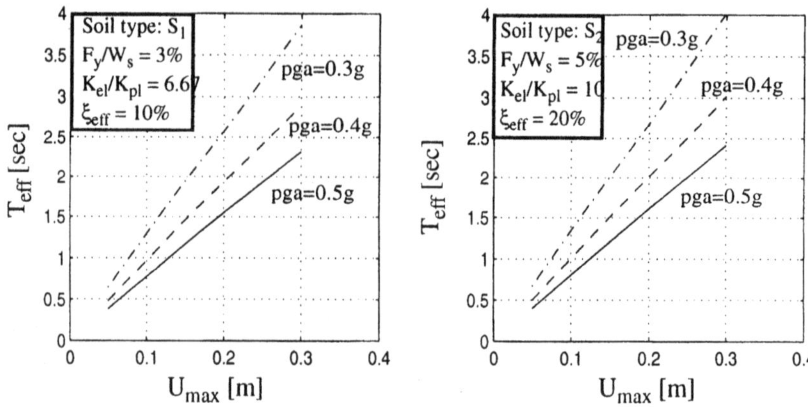

Figure 8.2: Design charts for required effective period T_{eff} based on U_{max}.

Then, the horizontal frequency of the structure f_h may be computed as

$$f_h = \frac{1}{T_{eff}} \tag{8.13}$$

8.6.4 Determination of plan size dimensions of isolators

The plan size dimensions of the isolators must be determined by the required vertical capacity, using the maximum axial forces on each bearing under various load combinations. The easiest way to compute the required area of each isolator A_i is to use the maximum axial forces from the maximum gravity loads combination. In addition, when the shape of the structure indicates the plausibility of occurrence of substantial additional axial forces, from overturning moments due to seismic loads, their effects may be approximately taken into account by their equivalent static application. When maximum axial loads from seismic combinations are used to determine the plan size dimensions of isolators, the effective area A_{eff} may be used instead of the actual area A, to take into account the deformed shape of the bearing. The effective area A_{eff} has been defined in Chapter 4 and is typically of the order of 0.6 to 0.8 of the area A. The vertical load capacity of an elastomeric bearing is determined by the plan dimensions of the bearing and the internal layer thickness. In addition, the displacement of the top layer with respect to the bottom reduces the area of the 'effective column', which may conservatively be considered to transfer vertical loads to the foundation.

The plan dimensions of the isolators must be chosen to ensure an approximately uniform distribution of vertical stresses well below the allowable compressive stress. As mentioned in Chapter 4, the plan area and the primary shape factor S_1, and consequently the internal layer rubber thickness, influence the vertical stiffness of the bearing f_v. This must be sufficiently large to avoid rocking of the superstructure. For most buildings, f_v is in the range of 8-12 Hz (Kelly [58] [59]). The ratio of the vertical to the horizontal frequency may be expressed using Equations (4.20) and (7.6) and may be simplified as follows:

$$\frac{f_v}{f_h} = \frac{\frac{1}{2\pi}\sqrt{K_v/W_s}}{\frac{1}{2\pi}\sqrt{K_h/W_s}} = \sqrt{\frac{K_v}{K_h}} = \sqrt{\frac{(6\ G\ S_1^2 + 3\ G)\ A_r/h_r}{G\ A_r/h_r}} \qquad (8.14)$$

$$fi\ S_1 = \sqrt{\frac{(f_v/f_h)^2 - 3}{6}}$$

When typical values of the ratio f_v/f_h are used, the above expression may be simplified to the following expression which may anyway be determined using Equation (4.18) instead of (4.20).

$$\frac{f_v}{f_h} = \sqrt{\frac{K_v}{K_h}} = \sqrt{\frac{6\ G\ S_1^2\ A_r/h_r}{G\ A_r/h_r}}\ fi\ S_1 = 0.408\ \frac{f_v}{f_h} \qquad (8.15)$$

Once the desired vertical frequency has been determined, therefore, the primary shape factor may in turn be computed. This factor is necessary in the determination of the plan dimensions of the bearings as it effects the bearings' vertical capacity.

The seismic effect on the maximum axial force on each bearing may be approximately taken into account, using equivalent static seismic forces combined with service loads. Subsequently, the superstructure may be analyzed as fixed supported for additional seismic loading combinations composed of service gravity loads and seismic loads in each horizontal direction. These loadings are based on the uniform distribution of the design base shear force F_{max} on the superstructure. The following expression may be used to compute F_{max}:

$$F_{max} = F_y \left(1 - \frac{K_{pl}}{K_{el}}\right) + K_{pl}\, U_{max} \tag{8.16}$$

As already mentioned, the distribution of the inertia seismic forces on an isolated structure have an approximately constant distribution, while that on a fixed supported structure linearly increases according to the shape of the fundamental mode. These forces must be applied in both the positive and the negative horizontal directions. Therefore, the equivalent static lateral forces applied to each mass m_j are given by

$$F_h^j = F_{max} \frac{m_j^s}{\sum m_j^s} \tag{8.17}$$

Using the maximum calculated axial loads at the bearing locations from all the load combinations and the allowable stresses at the bearings, with appropriate safety factors, the required plan size dimensions of each bearing are computed. The allowable compressive stress is the minimum of 6.89 MPa and 0.689 times the product of the shear modulus G and the primary shape factor S_1 (Roeder et al. [98]).

$$\sigma_{all}[MPa] = (min(6.895[MPa], 0.6895\ G[MPa]\ S_1))/SF \tag{8.18}$$

A safety factor SF may be used for possible variations of the compressive strength. Typical values of the primary factor S_1 are in the range of 10 to 20.

When circular bearings are used, and similarly for square bearings, we may obtain the required diameter as a function of the corresponding to the bearing maximum applied vertical load W^i_{max}.

$$D_i[m] = \sqrt{\frac{4\ W^i_{max}[MN]}{\pi\ \sigma_{all}[MPa]}} \tag{8.19}$$

As previously mentioned, for the maximum axial loads from seismic loading combinations, the reduction of the area to the effective area A_{eff} must be taken into account. This may be achieved by multiplying of the above derived dimension by the square root of the ratio of A to A_{eff}. A typical A_{eff}/A ratio of 0.7 will lead to an increase of 20% of the above-computed required diameter. Once the diameter has been determined, it may be used, together with the primary shape factor, to determine the internal rubber sheet thickness t_r. Typical values of the latter are within 5-15 mm.

$$t_r = \frac{D}{4\ S_1} \tag{8.20}$$

8.6.5 Determination of required total rubber thickness

The total rubber thickness must be determined to provide the required flexibility to the structure. Increasing the total height of the rubber increases the inserted flexibility; in other words it decreases the stiffness of the isolation system. Having computed the post-elastic stiffness K^i_{pl} and the area A_r^i of each bearing, the total height of the rubber h_r^i may be determined by the following formula:

$$h_r^i = \frac{G_r\ A_r^i}{K^i_{pl}} \tag{8.21}$$

Then, the number of required rubber sheets n_r for each bearing may be determined using the internal rubber sheet thickness t_r and the total height of the rubber h_r^i. The number of steel internal plates n_s is also derived from n_r.

$$n_r = \frac{h_r}{t_r} \quad \text{and} \quad n_s = n_r - 1 \tag{8.22}$$

The thickness of the internal steel plates t_s, typically of the order of 3 mm, must be selected. The total height of the rubber h_{tot} of each bearing may then be computed using the previously determined dimensions and the thickness t_p of the top and bottom plates, which is typically 20-40 mm.

$$h_{tot} = n_r\ t_r + n_s\ t_s + 2\ t_p \tag{8.23}$$

8.6.6 Design corrections and adjustments

This stage requires engineering judgement since it involves adjustments of the selected parameters to obtain an acceptable preliminary design. There are several issues to be checked. First, bearings must be checked to ensure that the maximum axial loads do not exceed the critical-for-buckling loading. In Chapter 4, the following expression has been derived for the critical load:

$$P_{cr} \approx \frac{0.134\ G_r\ D_r^5}{h_r\ t_r^2} \tag{8.24}$$

A safety factor SF_b for buckling must be used, i.e.

$$P_{max}^i \leq \frac{0.134\ G_r\ D_r^{i\,5}}{SF_b\ h_r^i\ t_r^{i\,2}} \tag{8.25}$$

Then the selected dimensions must be checked to ensure that they fall within acceptable ranges; otherwise they must be modified. Shear strains of the isolators γ_i must be computed from the total rubber thickness h_r^i of each bearing, and the maximum displacement U^i_{max}, and must be subsequently checked to ensure that they are below the allowable strains γ_{all}, i.e.

$$\gamma_i = \frac{U_{max}}{h_r} \leq \gamma_{all} \tag{8.26}$$

As previously mentioned, the diameter of the lead plug of each bearing d_l^i must also be checked to ensure that it lies within the range of one-third to one-sixth of the bearing diameter D_r. and is less than two-thirds of the total rubber height h_{tot}.

$$D_r/6 < d_l < D_r/3 \quad \text{and} \quad 1.5\ d_l \leq h_{tot} \tag{8.27}$$

When one tries to adjust the dimensions of the lead plugs of some bearings, it is possible that these dimensions may be too small. In such cases, it is preferable to have the bearings without lead plugs. The latter may instead be distributed only to external columns to satisfy the initial stiffness requirement and increase the torsional resistance.

With regard to the prior-to-yielding stiffness of the isolation system, one must first determine the stiffness of each bearing K^i_{el}. The following expression may be used:

$$K^i_{el} = \frac{G_l\ A^i_l}{h^i_{tot}} + \frac{G_r\ A^i_r}{h^i_r} \tag{8.28}$$

The system's overall prior-to and post-yield stiffnesses K_{el} and K_{pl}, respectively, are then calculated by summing the individual stiffnesses, i.e.

$$K_{el} = \sum K^i_{el} \qquad K_{pl} = \sum K^i_{pl} \tag{8.29}$$

The center of mass of the superstructure must be determined in order to check eccentricities. This is achieved using the axial column forces under the service loads and an auxiliary coordinate system, i.e.

$$X_{SM} = \frac{\sum W^i_s\ x^i}{\sum W^i_s}\ , \quad Y_{SM} = \frac{\sum W^i_s\ y^i}{\sum W^i_s} \tag{8.30}$$

Then the eccentricities are determined by a cartesian coordinate system with its origin at the center of mass *CM*, as shown in Figure 8.3.

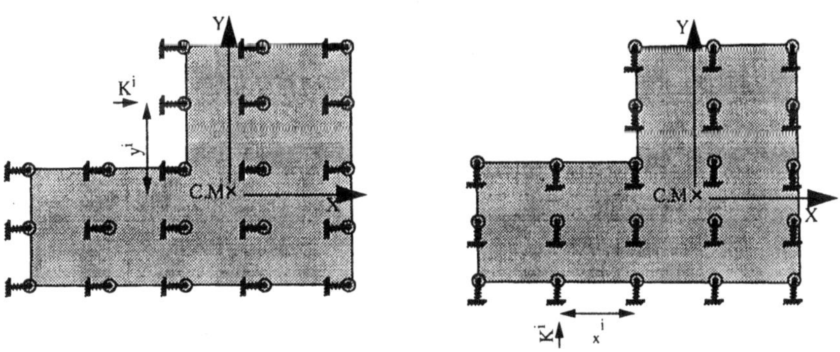

Figure 8.3: Calculation of eccentricities.

The eccentricities e_{el}^x and e_{el}^y, between the center of stiffness prior to yielding and the center of mass, may be computed as follows:

$$e^x_{el} = \frac{\sum K^i_{el}\ x^i}{\sum K^i_{el}}\ , \quad e^y_{el} = \frac{\sum K^i_{el}\ y^i}{\sum K^i_{el}} \tag{8.31}$$

Similarly, the eccentricities e_{pl}^x and e_{pl}^y are

$$e_{pl}^x = \frac{\sum K_{pl}^i x^i}{\sum K_{pl}^i} , \qquad e_{pl}^y = \frac{\sum K_{pl}^i y^i}{\sum K_{pl}^i} \tag{8.32}$$

From the preliminary procedure discussed above we deduce that the designer must experiment with different values of the above-mentioned critical parameters, and the dimensions of the components of the isolation system in particular. A trial and error process is necessary ultimately to satisfy the different design requirements. Once this procedure has been completed, the designer will have initial values of these parameters which are needed in the final and more detailed analysis.

8.7 Issues of Concern at the Final Analysis Stage

A three-dimensional time history dynamic analysis, for several appropriately selected accelograms, must be performed in the final stage of the design in order to provide highly accurate results, which cannot be obtained with simpler models. This analysis should take into account the vertical displacements, possible uplifts of the isolators, horizontal distribution of seismic forces, especially at the isolation level, torsional effects due to eccentricities and, in general, local results for individual structural elements and isolators. Finally, a non-linear dynamic analysis is necessary when the isolation system is highly non-linear, explicitly to take into account the hysteretic energy dissipation mechanism, and to investigate the possibility of higher-mode effects. The results of this more detailed analysis will indicate necessary adjustments to the isolation system.

The fundamental frequencies of seismically isolated structures in the vertical direction may fall into the frequency range of high vertical ground accelerations, especially in regions close to an active fault. When strong vertical ground motion components are expected, it would be useful to study the effects of a vertical excitation on the isolated structure. Therefore, time history dynamic analysis with simultaneous application of horizontal and vertical excitations should be used to take into account the increased seismic demands due to the vertical ground accelerations. The selection of the accelograms must be based on site characteristics, design earthquake magnitude, expected predominant period and distance from active faults. Thus, the availability of a seismic database with strong excitations would be useful for the appropriate selection of ground motion time histories. The dominant period of the ground motion tends to increase with increased layer depth and decreased soil stiffness. Stiff soils amplify the high-frequency components of ground motion, therefore, while deep soft soils amplify the low-frequency ones. The high-frequency components have a tendency to attenuate faster than the low ones, resulting in a dominant period which lengthens with the epicentral distance. The depth of the seismic fault also influences the expected peak ground acceleration. Long components of the earthquake excitation may also be present due to near-fault effects, or when the epicentral distance is large and seismic waves propagate through deep soft layers of soil. In the case where pertinent information is not available in the earthquake database, excitations may be scaled by an appropriate time-scale factor corresponding to the expected dominant period. The selection of the appropriate scaling of excitations is of particular importance for seismically isolated structures. This is because when accelograms with a short predominant period are used in the analysis, and long components of ground motion are expected in contrast, the effectiveness of seismic isolation is overestimated. Yielding of the superstructure must be avoided during the design earthquake; for the maximum credible earthquake, limited ductile inelastic deformations in the superstructure may be accepted only when they are unavoidable. The different effect of yielding on a seismically isolated structure from that on a fixed supported one has been discussed in Chapter 7 and must be taken into account.

When a three-dimensional analysis is performed, the isolated structure is more accurately modeled and more reliable results are obtained. When a nonlinear isolation system is used, the hysteretic energy dissipation may now be directly considered using the nonlinear behavior of the isolators. The higher-mode effects due to the bilinear behavior of the isolation system may also be studied to ensure the prevention of high floor accelerations. Torsional effects due to accidental eccentricities may also be more accurately assessed. It must be ensured that during the design earthquake the maximum computed lateral displacement is well below the distance provided by the seismic gap. In the maximum credible earthquake, the corresponding displacement must again be checked to ensure that it is still below the seismic gap. A safety margin must be included to allow additional displacements due to torsional effects since in practice they are almost never completely avoided. From the maximum displacement and the bearings' dimensions, the shear deformations may be computed and checked to be within allowable limits. Necessary adjustments must be made according to guidelines corresponding to those mentioned at the preliminary design procedure.

Axial forces must be determined and checked to ensure that the isolators will not be damaged, either due to buckling resulting from compressive axial stresses, or due to extensive tensile stresses. In Chapter 4 the critical-for-buckling, loading has been determined. It may be used here as it has also been used in the preliminary design procedure. The tensile stresses may result in overturning moments on columns with low gravity loads. When the axial stresses on the bearings are checked for seismic loads, the reduced effective area A_{eff} must be used. Tensile stresses may be introduced to elastomeric bearings due to uplift forces, which cancel the gravity loads. The possibility of isolators uplift must be checked for seismic loads and reduced dead loads. High tensile stresses may generate flaws in the elastomer which may progressively grow with reversed loading, leading to failure. Appropriate location and design of the bearings is required to reduce or avoid tensile stresses. In addition, construction details may prevent the transmission of tensile forces. Finally, a detailed nonlinear analysis at the final analysis stage may be used to take into account the nonlinear effect of axial load on horizontal stiffness, and the reduction of the critical load with increasing horizontal displacements.

The superstructure must be designed to sustain at least the design earthquake with no damage, i.e. without inelastic deformations. The current ductility requirements used for conventionally designed earthquake-resistant structures must not be relaxed for seismically isolated structures, even though ductility requirements are limited for them. As previously mentioned, the reasons are the possibility of vertical and long period components, the near-fault effects and the current limited experience with seismic isolation. In addition, the reduction of the effectiveness of seismic isolation, when the superstructure experiences inelastic deformation due to its softening, may be counterbalanced by additional energy dissipation when ductility capacity is available. During the maximum credible earthquake the superstructure must exhibit limited or no inelastic deformations. The isolation system must be less conservatively checked for this loading, to withstand the excitation without failure. Also, the maximum displacement must not exceed the provided seismic gap. The possibility of such excess displacement may be accounted for by introducing stoppers and bumpers, as described in

Chapter 6. Whenever necessary, adjustments on the preliminary selected design must be made to satisfy the design requirements. Once the design has been significantly modified, a new 3-D analysis must be performed and the new response checked again.

Chapter 9

The Future of Seismic Isolation

9.1 Summary

Seismic isolation has been presented in this book as a performance-based design method to account for earthquake excitations in bridges and low- to medium-rise buildings. Seismic isolation is emerging as a new trend in earthquake resistance design with great potential for the future. A seismically isolated structure experiences reduced seismic forces and accelerations and moves essentially as a rigid body, preventing damage due to deformations. This is achieved by mounting the structure on horizontally flexible devices, known as seismic isolators, which make the structure move in a much slower and smoother fashion, avoiding resonance. In addition, the substantially reduced seismic energy imparted to the structure is dissipated by the isolators, which are properly designed for such a purpose, rather than through inelastic deformations of the structure, which is what causes damage to its structural and non-structural components. Seismic isolation can be compared to the suspension system of a vehicle, which cushions the passengers, contents and the vehicle itself while traveling over a bumpy road.

First, the limitations of conventional earthquake resistance design and the resulting damage, which is no longer acceptable, has motivated the exploration of the characteristics and the potential of this design method. Conventional earthquake resistance design has been described in Chapter 2. Its limitations and disadvantages, particularly for stiff structures, have been discussed. The focus of conventional design on the prevention of structural collapse and the avoidance of casualties has been emphasized. The disadvantage of a conventionally designed structure's being in resonance with the usual ground motions has also been demonstrated. The fact that this method relies on the ability of the structure, when properly designed and constructed, to provide high levels of ductility, has been pointed out, as well as the fact that damage under strong excitations is anticipated, as a means of absorbing energy and avoiding structural collapse. Essentially, conventional earthquake resistance design, which is life-safety oriented, can be used to make a structure earthquake resistant, thereby avoiding its collapse, but cannot make a structure earthquake proof from damage. Subsequently, it has been argued that since the previously accepted damage level is no longer tolerated, a different design methodology is needed in designing earthquake-resistant structures. In recent years, there has been a greater demand for buildings designed not only to undergo a powerful earthquake without structural collapse, not only in order to avoid human casualties, but also to protect material assets, such as the structure itself and its contents, and to prevent loss of functionality of the building.

Seismic isolation has next been presented as an alternative method of design and construction to address the issue of damage. Its main feature, decoupling the structure from the destructive ground motion, has been explained, as well as its success in reducing induced seismic forces on the superstructure. Seismically isolating a structure reduces the induced seismic forces to magnitudes that do not cause inelastic deformations and, thus, damage to its structural and non-structural components. In addition, floor accelerations are substantially reduced, which prevents damage to the contents of a building.

In Chapter 3, the basic components, advantages, constraints and limitations of seismic isolation have been described. A comparative study of a system with two boundary conditions, namely fixed supported and seismically isolated, has been performed to demonstrate the better performance of seismic isolation. In particular, the reduced inter-story deflections and floor accelerations, at a cost of large relative displacement at the isolation level, have been reported. In addition to the protection of the structure and its contents, avoidance of business interruption, which is vital to the local and global economy, and loss of functionality of essential public facilities, which must remain operational after a strong earthquake, have been considered as important reasons for the employment of the method. Factors which enable the use of seismic isolation have been presented, as well as economic aspects related to its practical implementation for both new construction and the seismic upgrading of existing structures. Finally, a brief account has been given of the reasons why seismic isolation is not yet extensively used.

Commonly used seismic isolation systems have been described in Chapter 4. The discussion has focused on elastomeric bearings, LRBs, HDRBs and LHDRBs in particular. The potential of a hybrid system, consisting of high-damping rubber with a lead plug, has been assessed. This system has both initial rigidity and a continuous energy dissipation mechanism which is less temperature-dependent than that of an HDRB system. In addition, it allows lower prior-to-yielding stiffness which provides a higher initial degree of isolation and lower high-mode effects, due to smoother stiffness changes. The use of high-damping rubber bearings with no lead plugs, under internal columns, has been identified as a means of increasing torsional resistance while reducing the prior-to-yielding stiffness. The mechanical and physical characteristics of elastomeric bearings have then been presented, followed by an investigation of several issues useful in the design of isolated structures. Expressions for the vertical stiffness, buckling load and effective column area of elastomeric bearings have been derived and have subsequently been used in the proposed preliminary design procedure.

A brief account of the current status of seismic isolation in countries where it is extensively used has been given in Chapter 5. Selected examples of seismically isolated structures have been presented. Practical applications of seismic isolation in the United States, Japan, New Zealand and Italy have been described. Emphasis has been placed on the observed performance of those structures which have already been subjected to earthquake excitations. The observed and sometimes recorded performance of seismically isolated structures demonstrated the effectiveness of the method in mitigating seismic hazards. The accelerations of the structural masses have consistently been less than the peak ground accelerations, while the masses of fixed supported structures, almost always undergo accelerations that are amplifications of the peak ground accelerations. The superior performance of these structures has demonstrated the capabilities and great advantages of seismic isolation, motivating its wider use.

Next, practical issues and problems arising when seismic isolation is employed have been identified, and ways to address them have been proposed, in Chapter 6. First, the configuration of a seismic isolation system in buildings has been described as well as practical problems and issues that may arise during the implementation of seismic isolation. Many of the issues that are considered for seismically isolated buildings are also valid for other types of seismically isolated structures. Then, practical considerations and issues for seismically isolated bridges have been identified and discussed. It has been emphasized that seismic isolation allows not only the lowering of the seismic forces induced on a bridge structure, but also the manipulation of the distribution of the seismic and non-seismic lateral forces in an optimal way that benefits the overall bridge design. Proper dimensioning of seismic isolators, such as plan size of the bearings, lead plug diameter, and total rubber thickness in the case of lead rubber bearings, can control the distribution of lateral forces, directing them away from substructures that may not have the necessary strength and ductility to resist them. Problematic distributions of lateral loads due to unavoidable variations in height can easily be avoided using properly dimensioned seismic isolators.

Chapter 6 continues with a presentation of physical maintenance requirements that must be met upon the installation of a seismic isolation system. Among these, the importance of the clearance around the structure and the possible necessity for mechanisms such as stoppers and bumpers, to constrain the relative displacement of the superstructure and avoid collisions with adjacent structures, have been emphasized. Finally, code requirements for seismic isolation, particularly in the United States, have been mentioned.

The structural modeling and analysis of seismically isolated structures have been investigated in Chapter 7. A modal analysis of a seismically isolated structure has first been performed, to demonstrate the dominance of the fundamental mode of the structure on its response. The characteristics of linear isolation systems have been subsequently described and the effect of viscous damping has been assessed. As has been shown, by increasing the viscous damping, the displacements under an earthquake excitation are reduced, while the accelerations are reduced for damping only up to approximately 20-30% of the critical damping; for higher values the accelerations increase. It has, therefore, been deduced that an optimum viscous damping may be computed for the reduction of floor accelerations. A description of bilinear systems has been given, and a comparative study of a linearly and a bilinearly isolated structure has then been performed. The relative performance of the two systems and that of a hybrid system with both viscous and hysteretic damping have shown the better performance of the hybrid system. Next, the linearization of the bilinear behavior of an isolation system has been attempted. As shown, the values of the different parameters of concern, when such a linearization is made, vary significantly according to both the structural and the earthquake characteristics. Next, a study of the effect of non-linearities has indicated that a bilinear hysteretic system with moderate non-linearity and relatively low prior-to-yielding stiffness and yielding force, combined with viscous damping, may incorporate the advantages of both linear and non-linear systems. The effect of superstructure stiffness has also been discussed and it has been demonstrated that its increase affects positively the efficiency of seismic isolation, especially the reduction of interstory deflections. Finally, the issue of ductility has been discussed, emphasizing its influence on a seismically isolated and a fixed supported structure. Inelastic deformations in a fixed supported structure are beneficial, due to energy dissipation and softening of the structure which takes it away from the dangerous-for-resonance range. In contrast, in the case of a seismically isolated structure, the corresponding effects are contradictory. On the one hand, the inelastic deformations provide an additional energy dissipation mechanism, but on the other hand, the softening of the superstructure reduces the effectiveness of seismic isolation.

Finally, a preliminary design methodology for seismically isolated structures has been proposed in Chapter 8. It may be used to obtain parameter values which are necessary in the final and more detailed analysis. This procedure has been based on bilinear isolation systems, and elastomeric bearings with lead plugs in particular, although the primary issues addressed are pertinent to most systems. A feasibility study has been suggested at the first stage of the preliminary design, to investigate the plausibility and effectiveness of seismic isolation. The level of the expected earthquake excitation and in particular the peak ground acceleration must also be estimated. Then the superstructure configuration must be determined and analyzed for both dead and live loads. The next step is to select an appropriate isolation system, given the particular design requirements. Finally, the preliminary design of the isolation system must follow, which in turn involves several stages.

Due to the several interrelated parameters of concerned, the preliminary design of the isolation system may require a trial and error procedure which will ultimately lead to an optimum design. First, the maximum allowable displacement must be estimated, considering practical constraints. Then, the yield force of the isolation system must be determined according to the minor lateral service loads, and the diameter of the lead plugs at each bearing must be computed. The required post-yield stiffness may be computed from a design spectrum after estimation of the equivalent viscous damping of the isolation system, the nonlinearity ratio K_{el}/K_{pl} and site characteristics. The corresponding fundamental period and frequency of the isolated structure may then be computed. Subsequently, the plan dimensions of each bearing may be determined using the requirement for the maximum allowable compressive stress and the vertical stiffness to avoid rocking. Finally, the total rubber thickness is determined to provide the computed post-yield stiffness. Having obtained the primary shape factor, the thickness of the internal rubber sheets may be computed, and subsequently the number of the rubber layers. The thickness of the steel plates may also be selected. These values are used to determine the total thickness of the bearing. Adjustments and corrections must also be performed to ensure that the design requirements are satisfied. First, the axial forces on the bearings must be checked to ensure that they are well below buckling load. Then, shear strains may be computed and checked against the maximum allowable shear strains. Finally, the prior-to- and post-yielding stiffness must be checked to ensure the desired stiffness and avoid torsional effects. The procedure may also require several adjustments and corrections to obtain an appropriate design of the isolation system. Issues of concern in the final and more detailed analysis have also been discussed. The importance of appropriately selected and scaled ground motion has been emphasized, since it has significant effects on the response of the isolated structure. Although seismic forces and the required amount of ductility in a seismically isolated structure are substantially reduced, the current design details to ensure ductility capacity must not be relaxed, until there is sufficient knowledge of the behavior of seismically isolated structures, and a reliable estimate of the expected earthquake excitation, are obtained.

9.2 Future of Seismic Isolation

It is evident from this discussion that seismic isolation is an efficient design scheme which successfully addresses earthquake loadings, and not only provides safety but also prevents damage. Seismic isolation is particularly useful for low- to medium- rise buildings which happen to have their fundamental frequencies within the dangerous-for-resonance range of dominant earthquake frequencies. Critical facilities, such as emergency response centers, hospitals, fire-stations, utilities and communication centers, should remain operational even during the strongest earthquakes. The disruption of the operation of such essential, for the public interest, facilities may be prevented by using seismic isolation to enhance their earthquake capacity by reducing the seismic loads that they may experience during a powerful earthquake.

It is also very useful, and is probably the only currently available technology that can be used, to seismically upgrade historic structures or to protect very sensitive equipment and the valuable contents of a building. In the former case, seismic isolation can decrease the seismic vulnerability of an historic structure without impairing its distinctive architectural characteristics, by reducing the induced seismic forces. In the latter case, the contents of a structure are protected by reducing the inertia forces that they may experience during a strong earthquake excitation.

The seismic risk that many businesses run by not taking earthquake protection measures is, often, too serious to be ignored. The disruption of businesses and commercial organizations can be devastating for them, due to production and revenue loss, in addition to potential damage to equipment and tangible assets. Discontinuation of production and suspended operations may lead to loss of sales, customers and market share, which might have taken years to acquire. Seismic isolation can cost-effectively mitigate, or at least significantly reduce, such seismic risk.

However, seismic isolation is not applicable to all structures; therefore a feasibility study is imperative at the early stages of design to assess its effectiveness according to particular design requirements. In addition, different design requirements need different isolation systems; thus, an appropriate selection must be made based on knowledge and understanding of the characteristics, advantages and limitations of each system.

As with any other innovative design method, seismic isolation may require some time before it becomes established as a method for dealing with seismic loads. Currently, it is not widely accepted mainly due to engineers' conservatism in being reluctant to adopt innovative construction methods, particularly when there are no immediate cost savings. Employment of seismic isolation requires an additional cost, although in the long term it ensures savings due to reduced damage and uninterrupted function of the structure. Recent seismic events have shown the necessity for more stringent codes, imposing performance requirements in order to avoid damage and ensure safety. It is therefore expected that seismic isolation will be more extensively used in the future.

Authorities should also promote the use of seismic isolation, as well as any other method with enhanced seismic performance, since their application will result in reduced damage and corresponding economic savings. In addition to the modification of codes, which will favor, or at least facilitate, the use of seismic isolation, information on the advantages of the method should also be more widely available. Insurance companies might also provide special premiums for structures that are seismically isolated, due to their better performance.

Seismic codes aim to avoid structural collapse and resulting casualties. According to SEAOC, *"Recommendations primarily are intended to safeguard against major failures and loss of life, not to limit damage, maintain functions, or provide for easy repair."* However, damage and property loss, as well as the disruption of a business or the loss of functionality of a structure, are issues that also need to be addressed by the codes. These consequences, although less important than life safety protection, cannot any longer be accepted. Seismic isolation will then find itself in a beneficial position, compared to conventional earthquake resistance approaches, since it aims not only to prevent structural collapse but also to avoid structural, nonstructural and contents damage.

A factor that discourages owners of newly constructed buildings to use seismic isolation is the illusion that a structure designed using conventional earthquake resistance methods is "earthquake proof". The lack of information and of awareness that conventional earthquake resistance philosophy is based on the acceptance of high levels of damage under strong earthquakes misleads owners, who do not realize the risk they are taking with their property. When the public is properly informed of the design philosophy and the objectives of conventional earthquake design approaches, there may be increased interest in seismic isolation, especially in the business world. Disruption of the operation of the headquarters of a large corporation, or interruption of industrial production, may be so devastating that it motivates the use of seismic isolation to prevent rather than to cure the consequences of strong earthquakes.

The publicity that has been given to recent earthquakes and the resulting damage has made people more sensitive and concerned about earthquakes and their consequences. It is hoped that people will start to pay more attention to the earthquake resistance of their buildings, demanding better performance and being willing to pay the extra initial cost in order to safeguard lives and property, and thereby saving much more in the long run. The cost of repairing the damage caused by a powerful earthquake can be a significant proportion of the national gross product of a country, and devastating for its economy. It is urgent therefore that countries in seismically prone areas should encourage the use of advanced earthquake resistance methods, such as seismic isolation, to mitigate the consequences of anticipated powerful earthquakes.

Isolation systems must also be standardized to allow their easier use. Certified characteristics of seismic isolation systems should be readily available to structural engineers to allow consideration of the method in any earthquake resistance design procedure. Also, future civil engineers must be exposed to the method, its characteristics and purpose, during their studies.

Widespread use of seismic isolation may be possible only when it becomes a cost-efficient approach to mitigating earthquake hazards. Unfortunately, everything in our world is driven by economics, and seismic isolation, in most cases, has an initial overhead cost, which makes it less attractive than conventional earthquake resistance design methods. The cost of seismic isolation can be reduced by adopting less conservative codes for seismically isolated structures, since they are protected by the isolation system. It is likely that as more seismic isolation structures are constructed, the cost will reduce. The higher demand will allow the companies that manufacture seismic isolation systems to profit from increased numbers of sales while lowering their prices to a level which promotes the use of seismic isolation. This will benefit both the seismic isolation industry, which will become better established, and society, since the mitigation of earthquake hazards will be possible. In addition, it may be possible to utilize modern materials in order to develop new types of seismic isolators that will be less expensive to manufacture For example, the use of fiber-reinforced elastomeric bearings may allow significant cost reductions since the use of steel plates and the expensive vulcanization process can be avoided.

Finally, as the number of seismically isolated structures increases, more experience will be gained regarding the performance of the method and its benefits. More experience and data from the responses of seismically isolated structures during violent earthquakes may help to reduce present conservative reluctance to use this innovative approach. The establishment of seismic isolation may be just the starting point. Other innovative hazard mitigation approaches, such as passive and active control, may follow. Also, it may be possible that seismic isolation can be combined with a particular control scheme, for example one to account for wind loads, in order to address the issue of multiple natural loadings.

Although we can neither prevent earthquakes nor accurately predict them, we can prepare for them so that lives are saved and damage is minimized. Instead of excessively strengthening a structure in an attempt to avoid its collapse in a powerful earthquake while accepting structural, non-structural and contents damage, it may be wiser to prevent the excessive shaking of that structure by seismically isolating it.

Appendix A

Linear Elastic Systems

A.1 Single Degree of Freedom Systems

The motion of the foundation ground due to an earthquake excitation results in inertia forces acting on the structural masses. The response of many buildings is mainly characterized by their fundamental eigenmode. The response of a seismically isolated building is dominated even more significantly by its fundamental eigenmode, considering the difference between the stiffness at the superstructure level and that at the isolation level, which results in rigid body motion.

Consider the case of a damped single degree of freedom system shown in Figure A.1.

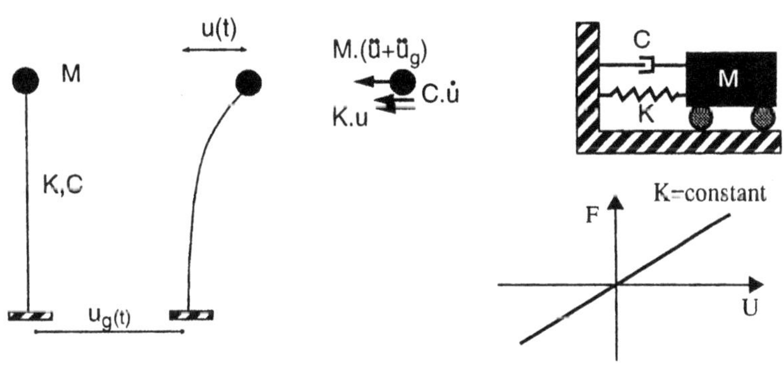

Figure A.1: Single DOF system subjected to support ground motion.

We apply Newton's law to the free body diagram in Figure A.1 and use D'Alembert's principle to obtain the governing equation of motion:

$$M\ \ddot{u}(t) + C\ \dot{u}(t) + K\ u(t) = -M\ \ddot{u}_g(t) \tag{A.1}$$

where M, C and K are the mass, damping and stiffness of the structure, respectively, $\ddot{u}(t)$, $\dot{u}(t)$ and $u(t)$ are the mass acceleration, velocity and displacement, and $\ddot{u}_g(t)$ is the ground acceleration.

For a specific ground excitation, the equation of motion may be solved using different methods. The Newmark α–δ method, an implicit, direct integration method has been selected. In the special case of the method which is based on the trapezoidal rule, α and δ have the values 0.25 and 0.50, respectively. Then, the method assumes that the velocity and displacement at time $(t+\Delta t)$ are given by

$$u(t+\Delta t) = u(t) + \frac{\dot{u}(t+\Delta t) + \dot{u}(t)}{2}\,\Delta t \qquad \text{(A.2)}$$

$$\dot{u}(t+\Delta t) = \dot{u}(t) + \frac{\ddot{u}(t+\Delta t) + \ddot{u}(t)}{2}\,\Delta t \qquad \text{(A.3)}$$

Substitute (A.3) in (A.2) to obtain

$$u(t+\Delta t) = u(t) + \dot{u}(t)\,\Delta t + 0.25\,\Delta t^2\,(\ddot{u}(t) + \ddot{u}(t+\Delta t)) \qquad \text{(A.4)}$$

and finally

$$\ddot{u}(t+\Delta t) = \frac{4}{\Delta t^2}\,(u(t+\Delta t) - u(t) - \Delta t\,\dot{u}(t) - 0.25\,\Delta t^2\,\ddot{u}(t)) \qquad \text{(A.5)}$$

The resulting expression for the acceleration is substituted in (A.3) to obtain

$$\dot{u}(t+\Delta t) = -\dot{u}(t) + 2\,\frac{(u(t+\Delta t) - u(t))}{\Delta t} \qquad \text{(A.6)}$$

Finally, the last three expressions (A.4) to (A.6) are used in the equation of motion to solve for the displacement u at time $(t+\Delta t)$; which is given by

$$u(t+\Delta t) = \frac{M\left\{\dfrac{4\,u(t)}{\Delta t^2} + \dfrac{4\,\dot{u}(t)}{\Delta t} + \ddot{u}(t) - \ddot{u}_g(t+\Delta t)\right\}}{\left(\dfrac{4\,M}{\Delta t^2} + \dfrac{2\,C}{\Delta t} + K\right)} + \qquad \text{(A.7)}$$

$$C\left\{\dot{u}(t) + \frac{2\,u(t)}{\Delta t}\right\} \Big/ \left(\frac{4\,M}{\Delta t^2} + \frac{2\,C}{\Delta t} + K\right)$$

Once this has been obtained under appropriate initial conditions, the displacement, velocity and acceleration of the system may be obtained iteratively, using Equations A.7, A.3 and A.5, respectively.

A.2 Multi Degree of Freedom Systems

Most low- to medium-rise buildings may be modeled as shear buildings, assuming that their floors are sufficiently rigid, so that they do not allow rotation about the horizontal axis, and that the columns are axially non-deformable. Consequently, the mass of the structure may be treated as lumped at the floor levels, having only three dynamic degrees of freedom per floor (two horizontal translations and a rotation about a vertical axis). The 'shear building' approximation assumes that the whole building deforms as a shear beam, although the actual 'internal' deformations are due mainly to bending. In addition, if we neglect eccentricities which result in torsional effects, and consider the earthquake excitation to be in one direction only, the building may be modeled by lumped masses with one translational degree of freedom per floor.

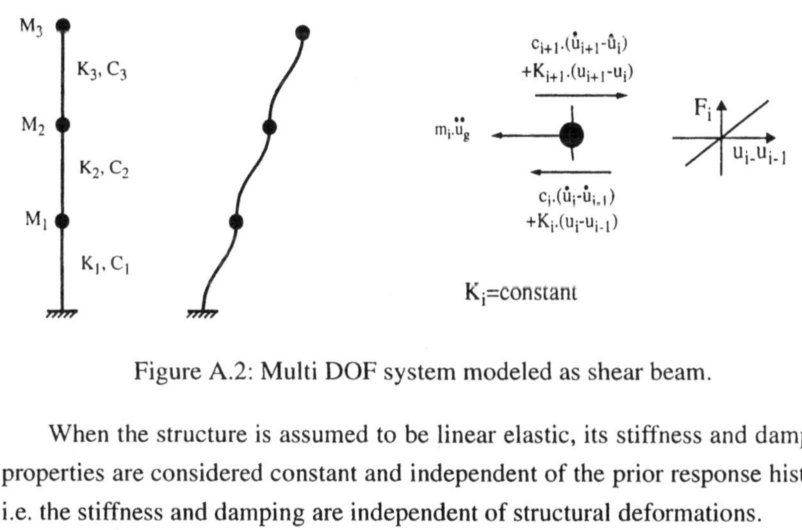

Figure A.2: Multi DOF system modeled as shear beam.

When the structure is assumed to be linear elastic, its stiffness and damping properties are considered constant and independent of the prior response history, i.e. the stiffness and damping are independent of structural deformations.

The equation of equilibrium for a mass M_i may be written as

$$M\ \ddot{u}_{tot} + C_i\ (\dot{u}_i - \dot{u}_{i-1}) - C_{i+1}\ (\dot{u}_{i+1} - \dot{u}_i) \qquad \text{(A.8)}$$

$$+ K_i\ (u_i - u_{i-1}) - K_{i+1}\ (u_{i+1} - u_i) = 0$$

where the total displacement is equal to the sum of the relative and ground displacements, i.e. $(u_{tot} = u + u_g)$.

The above equation holds for all masses M_i. We can therefore write the equilibrium equation of the system in matrix form:

$$[M]\,[\ddot{U}_{tot}(t)] + [C]\,[\dot{U}(t)] + [K]\,[U(t)] = [0] \tag{A.9}$$

where $[U_{tot}] = [U(t+\Delta t)] + [e]\,u_g(t+\Delta t)$, with $[e]$ a unit vector. Thus,

$$[M]\,[\ddot{U}(t)] + [C]\,[\dot{U}(t)] + [K]\,[U(t)] = -[M]\,[\ddot{U}_g(t)] \tag{A.10}$$

The above system of differential equations may be solved using one of several numerical methods. When the system is linear elastic, the mode superposition analysis method may be used, since superposition of individual mode responses is allowed. When the problem involves non-linearities, a time step-by-step history analysis, such as the Newmark method, must be performed.

For the particular case of a three-mass model, shown in Figure A.3, the system matrices are:

$$[M] = \begin{bmatrix} m_1 & 0 & 0 \\ 0 & m_2 & 0 \\ 0 & 0 & m_3 \end{bmatrix} \tag{A.11}$$

$$[C] = \begin{bmatrix} (c_1+c_2) & -c_2 & 0 \\ -c_2 & (c_2+c_3) & -c_3 \\ 0 & -c_3 & c_3 \end{bmatrix} \tag{A.12}$$

$$[K] = \begin{bmatrix} (k_1+k_2) & -k_2 & 0 \\ -k_2 & (k_2+k_3) & -k_3 \\ 0 & -k_3 & k_3 \end{bmatrix} \tag{A.13}$$

A.2.1 Modal analysis

The response of a multi-DOF structure may be computed as the sum of the responses of independent single-DOF systems, corresponding to the eigenmodes of the structure. The system of differential equations of motion may be uncoupled by means of a transformation of coordinates which incorporates the orthogonality property of modal shapes.

$$\lfloor \phi_i \rfloor \lfloor M \rfloor \lfloor \phi_j \rfloor = \lfloor \phi_i \rfloor \lfloor K \rfloor \lfloor \phi_j \rfloor = 0 \text{ when } i \neq j$$

The eigenmodes of an undamped system may be computed for the free vibration of the system. In such a case the equation of motion of the system is

$$\lfloor M \rfloor \lfloor \ddot{U} \rfloor + \lfloor K \rfloor \lfloor U \rfloor = \lfloor 0 \rfloor$$

Assuming that the displacements may be expressed as $\lfloor U \rfloor = \lfloor \phi \rfloor \; sin(\omega t)$

$$\lfloor K \rfloor \lfloor \phi \rfloor = \omega^2 \lfloor M \rfloor \lfloor \phi \rfloor$$

The eigenproblem may be solved numerically to obtain the N natural frequencies and mode shapes. Inverse iteration is used, combined with 'sweeping' for convergence purposes. The proccss is illustrated below:

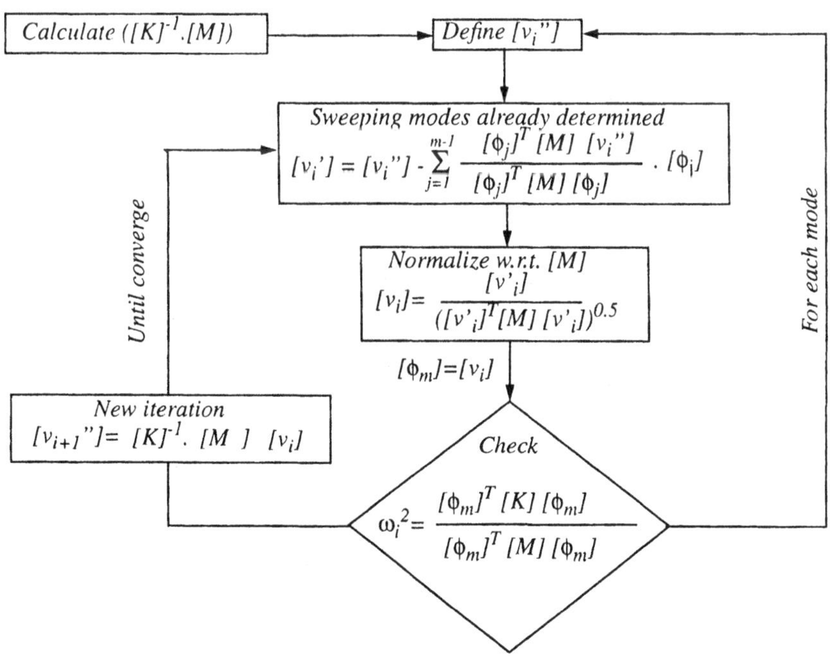

Figure A.3: Inverse iteration procedure.

Introducing the transformation $\lfloor U \rfloor = [\Phi] \lfloor y \rfloor$, the system of equations becomes

$$[M]\,[\Phi]\,\lfloor \ddot{y} \rfloor + [C]\,[\Phi]\,\lfloor \dot{y} \rfloor + [K]\,[\Phi]\,\lfloor y \rfloor = -[M]\,\lfloor e \rfloor\,\ddot{u}_g(t + \Delta t) \qquad \text{(A.14)}$$

Premultiplying by the transpose of each mode ϕ_i, we obtain

$$\lfloor \phi_i \rfloor^T [M]\,[\Phi]\,\lfloor \ddot{y} \rfloor + \lfloor \phi_i \rfloor^T [C]\,[\Phi]\,\lfloor \dot{y} \rfloor + \lfloor \phi_i \rfloor^T [K]\,[\Phi]\,\lfloor y \rfloor = -\lfloor \phi_i \rfloor^T [M]\,\lfloor e \rfloor\,\ddot{u}_g(t + \Delta t) \qquad \text{(A.15)}$$

Considering the orthogonality property of modes, and assuming that the system is damped proportionally to the mass and stiffness, the $(N \times N)$ system of equations may be uncoupled in a set of N-independent differential equations of the form

$$\lfloor \phi_i \rfloor^T [M]\,\lfloor \phi_i \rfloor\,\lfloor \ddot{y} \rfloor + \lfloor \phi_i \rfloor^T [C]\,\lfloor \phi_i \rfloor\,\lfloor \dot{y} \rfloor + \lfloor \phi_i \rfloor^T [K]\,\lfloor \phi_i \rfloor\,\lfloor y \rfloor = -\lfloor \phi_i \rfloor^T [M]\,\lfloor e \rfloor\,\ddot{u}_g(t) \qquad \text{(A.16)}$$

which may be simplified as

$$m_i\,\ddot{y}_i + c_i\,\dot{y}_i + k_i\,y_i = -\lfloor \phi_i \rfloor^T [M]\,\lfloor e \rfloor\,\ddot{u}_g(t) \qquad \text{(A.17)}$$

where $m_i = \lfloor \phi_i \rfloor^T [M]\,\lfloor \phi_i \rfloor$ is the modal mass, $c_i = \lfloor \phi_i \rfloor^T [C]\,\lfloor \phi_i \rfloor$ the modal damping, and $k_i = \lfloor \phi_i \rfloor^T [K]\,\lfloor \phi_i \rfloor$ the modal stiffness.

By dividing each of the N-differential equations by m_i, we obtain:

$$\ddot{y}_i + 2\,\xi_i\,\omega_i\,\dot{y}_i + \omega_i^2\,y_i = -\Gamma_i\,\ddot{u}_g \qquad \text{(A.18)}$$

where $2\ \xi_i\ \omega_i = \dfrac{c_i}{m_i}$, $\omega_i^2 = \dfrac{k_i}{m_i}$, and $\Gamma_i = \dfrac{\lfloor \phi_i \rfloor^T \lfloor M \rfloor \lfloor e \rfloor}{m_i} = \dfrac{L_i}{m_i}$

We can simplify the derived equations further by introducing the transformation $y_i = \Gamma_i\ \psi_i$, which transforms the set of N equations to the same form as the differential equation of motion of a single DOF system subjected to an earthquake excitation $\ddot{u}_g(t)$:

$$\ddot{\psi}_i + 2\ \xi_i\ \omega_i\ \dot{\psi}_i + \omega_i^2\ \psi_i = -\ddot{u}_g \tag{A.19}$$

The resulting displacements, velocities and accelerations may be computed via modal superposition as

$$\lfloor U(t) \rfloor = \lfloor \Phi \rfloor\ \lfloor y(t) \rfloor = \sum_{i=1}^{N} \lfloor \phi_i \rfloor\ y_i(t) = \sum_{i=1}^{N} \lfloor \phi_i \rfloor\ \Gamma_i\ \psi_i(t) \tag{A.20}$$

$$\lfloor \dot{U}(t) \rfloor = \lfloor \Phi \rfloor\ \lfloor \dot{y}(t) \rfloor = \sum_{i=1}^{N} \lfloor \phi_i \rfloor\ \dot{y}_i(t) = \sum_{i=1}^{N} \lfloor \phi_i \rfloor\ \Gamma_i\ \dot{\psi}_i(t) \tag{A.21}$$

$$\lfloor \ddot{U}(t) \rfloor = \lfloor \Phi \rfloor\ \lfloor \ddot{y}(t) \rfloor = \sum_{i=1}^{N} \lfloor \phi_i \rfloor\ \ddot{y}_i(t) = \sum_{i=1}^{N} \lfloor \phi_i \rfloor\ \Gamma_i\ \ddot{\psi}_i(t) \tag{A.22}$$

$$\lfloor \ddot{U}_{total}(t) \rfloor = \lfloor \ddot{U}(t) \rfloor + \lfloor \ddot{U}_g(t) \rfloor = \sum_{i=1}^{N} \lfloor \phi_i \rfloor\ \Gamma_i\ \ddot{\psi}_i(t) + \lfloor \ddot{U}_g(t) \rfloor \tag{A.23}$$

The forces may then be computed from the displacements as

$$\lfloor F \rfloor = \lfloor K \rfloor\ \lfloor U \rfloor = \lfloor K \rfloor \sum_{i=1}^{N} \lfloor \phi_i \rfloor\ \Gamma_i\ \psi_i = \sum_{i=1}^{N} \Gamma_i\ \psi_i\ \omega_i^2\ \lfloor M \rfloor\ \lfloor \phi_i \rfloor \tag{A.24}$$

Finally, the base shear force can then be calculated as the sum of the horizontal forces acting on the structure.

$$Q = \sum F_i = \lfloor e \rfloor^T \lfloor F \rfloor = \tag{A.25}$$

$$\sum_{i=1}^{N} \Gamma_i \; \psi_i \; \omega_i^2 \; \lfloor e \rfloor^T \; \lfloor M \rfloor \; \lfloor \phi_i \rfloor = \sum_{i=1}^{N} \left(\frac{L_i^2}{m_i} \right) \psi_i \; \omega_i^2$$

A.2.2 Newmark method

The Newmark method may be used to perform a time history analysis of a multi-DOF system. For the case where α=0.25 and δ=0.50, the expressions for the displacement, velocity and acceleration vectors are the same as those for a single-DOF system, but in vector form. The governing equation of motion finally obtained is:

$$\lfloor M \rfloor \left(\frac{4}{\Delta t^2} \left(\lfloor U(t+\Delta t) \rfloor - \lfloor U(t) \rfloor - \lfloor \dot{U}t \rfloor \; \Delta t - 0.25 \; \Delta t^2 \; \lfloor \ddot{U}(t) \rfloor \right) \right) + \tag{A.1}$$

$$[C] \left(2 \; \frac{\left(\lfloor U(t+\Delta t) \rfloor - \lfloor U(t) \rfloor \right)}{\Delta t} - \lfloor \dot{U}t \rfloor \right) +$$

$$[K] \; \lfloor U(t+\Delta t) \rfloor = -\lfloor M \rfloor \; \lfloor e \rfloor \; \ddot{U}_g(t+\Delta t)$$

Equation A.26 can be solved for $U(t + \Delta t)$:

$$\lfloor U(t+\Delta t) \rfloor = \left\{ \frac{4}{\Delta t^2} \; \lfloor M \rfloor + \frac{2}{\Delta t} \; \lfloor C \rfloor + \lfloor K \rfloor \right\}^{-1} \tag{A.2}$$

$$\left\{ C \left(\lfloor \dot{U}t \rfloor + \frac{2}{\Delta t} \; \lfloor Ut \rfloor \right) - \lfloor M \rfloor \; \lfloor e \rfloor \; \ddot{U}_g(t+\Delta t) + \right.$$

$$\left. \lfloor M \rfloor \left(\frac{4}{\Delta t^2} \; \lfloor U(t) \rfloor + \frac{4}{\Delta t} \; \lfloor \dot{U}(t) \rfloor + \lfloor \ddot{U}(t) \rfloor \right) \right\}$$

Appendix B

Bilinear Systems

Many isolation systems are characterized by a highly nonlinear behavior and hysteretic dissipation of energy through cycles of loading and unloading. The most commonly used isolation systems, such as lead rubber bearings, behave bilinearly. The lumped mass models described in Appendix A may be easily modified to consider local non-linearities.

B.1 Single Degree of Freedom Systems

The equation of motion of a bilinear single DOF system is given by

$$M\ \ddot{u}(t) + C(t)\ \dot{u}(t) + K(t)\ u(t) = -M\ \ddot{u}_g(t) \tag{B.1}$$

where the stiffness K and the damping C of the single DOF system are no longer constant but depend on the previous response history of the system.

$$K = K(t) \text{ and } C = C(t)$$

The stiffness K at any time $t + \Delta t$ is based on the prior response history of the system and the magnitude of the force $F(t)$. The stiffness may be determined at each step, from which the damping may in turn be computed. Assume that a particular damping ratio ξ is given. The damping is expressed as

$$c(t + \Delta t) = 2\ \xi\ \omega\ M \tag{B.2}$$

$$\text{where } \omega = \sqrt{\frac{K(t + \Delta t)}{M}} \tag{B.3}$$

The displacement, velocity, acceleration and force are computed at each time step taking into account the yield force. Whenever the yield force level is exceeded, the post-yield stiffness $K_{isol,pl}$ is used, instead of the initial elastic stiffness $K_{isol,el}$. In addition, the loading path is reversed whenever there is a change in the sign of the velocity $\dot{u}(t)$. If the change in the sign of the velocity happens during a post-yield branch of the force-displacement diagram, then the initial elastic stiffness $K_{isol,el}$ is used, instead of the post-yield stiffness $K_{isol,pl}$. Following these simple rules the nonlinear loading and unloading paths are generated. The force $F(t+\Delta t)$ at time $t+\Delta t$ is determined based on the value of the force $F(t)$ at time t, the computed displacements $u(t)$ and $u(t+\Delta t)$, and the stiffness $K(t+\Delta t)$ during the Δt step, as

$$F(t+\Delta t) = K(t+\Delta t)\{u(t+\Delta t)-u(t)\} \quad \text{(B.4)}$$

In order to be able to identify the exact time instants at which either the yield force level is exceeded or the velocity sign changes, an algorithm, which uses a linear interpolation procedure, has been developed and implemented. In particular, the algorithm at each time instant t (Figure B.1(a)), while computing the response at time $t+\Delta t$, checks whether the force $F(t+\Delta t)$ exceeds the yield force F_y. In that case, i.e. when the force $F(t+\Delta t)$ exceeds, in absolute value, the corresponding yield force F_y, as shown in Figure B.1(b), a linear interpolation procedure is used to determine the exact time instant $t+\lambda\Delta t$, where $0 \leq \lambda \leq 1$, at which $F(t+\lambda\Delta t) = F_y(t+\lambda\Delta t)$. The algorithm employed is similar to the Regula-Falsi method which is used to solve nonlinear equations.

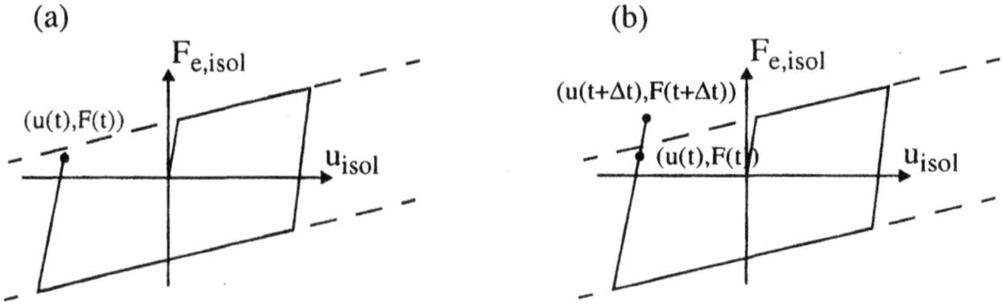

Figure B.1: Linear interpolation procedure to determine yield point.

In particular, when the computed force $F(t+\Delta t)$ exceeds the yield force F_y, which also depends on the current displacement, a linear interpolation between $F(t)$ and $F(t+\Delta t)$ is performed to estimate the time $t+\lambda_1\Delta t$ at which the force equals the yield force at that location. The ground acceleration at time $t + \lambda_1 \Delta t$ is estimated by linearly interpolating between the ground accelerations $a_g(t)$ and $a_g(t + \Delta t)$, Then, using the time step $\lambda_1 \Delta t$ and the ground acceleration $a_g(t + \lambda_1 \Delta t)$, the force $F(t + \lambda_1 \Delta t)$ at time $t + \lambda_1 \Delta t$ is computed and compared with $F_y(t + \lambda_1 \Delta t)$. Based on the difference between $F_y(t + \lambda_1 \Delta t)$ and $F(t + \lambda_1 \Delta t)$, a new linear interpolation is performed. Continuing this iterative procedure, after a few iterations the exact time instant $t + \lambda_i \Delta t$ at which $F(t + \lambda_i \Delta t) = F_y(t + \lambda_i \Delta t)$ is determined as shown in Figure B.2(a).

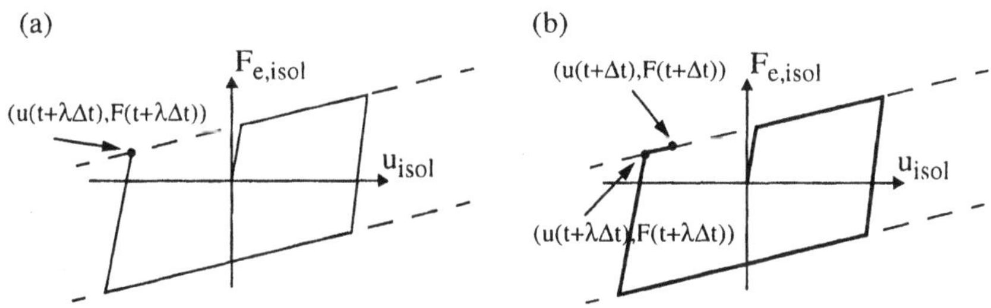

Figure B.2: Linear interpolation procedure to determine yield point (cont.).

Once the time of yielding has been determined, the time history analysis continues with the updated post-yield stiffness $K_{isol,pl}$ (Figure B.2(b)).

A similar procedure is used to determine the exact time instant at which the sign of velocity changes. For example, having determined the response at time t, as shown in Figure B.3(a), while computing the response at time $t + \Delta t$ the velocity $\dot{u}(t + \Delta t)$ is checked, to see whether it changes sign. If the velocity changes sign, a linear interpolation procedure, similar to the Regula-Falsi method above, is used to determine the exact time instant $t + \lambda \Delta t$, where $0 \le \lambda \le 1$, at which $\dot{u}(t + \lambda \Delta t) = 0$.

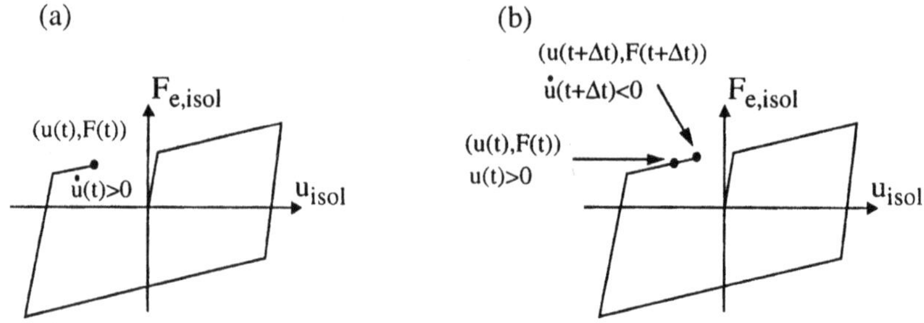

Figure B.3: Linear interpolation procedure to determine a zero velocity instant.

Linearly interpolating the ground accelerations $a_g(t)$ and $a_g(t + \Delta t)$, according to the values of $\dot{u}(t)$ and $\dot{u}(t + \Delta t)$, the ground acceleration at time $t + \lambda_1 \Delta t$ is estimated, as in Figure B.4(a). Then, using the time step $\lambda_1 \Delta t$ and the ground acceleration $a_g(t + \lambda_1 \Delta t)$, the velocity $\dot{u}(t + \lambda_1 \Delta t)$ at time $t + \lambda_1 \Delta t$ is computed and compared with zero. Based on the difference between $\dot{u}(t + \lambda_1 \Delta t)$ and zero, a new linear interpolation is performed. Continuing this iterative procedure, after a few iterations the exact time instant $t + \lambda_i \Delta t$ at which $\dot{u}(t + \lambda_i \Delta t) = 0$ is determined, as shown in Figure B.4(b).

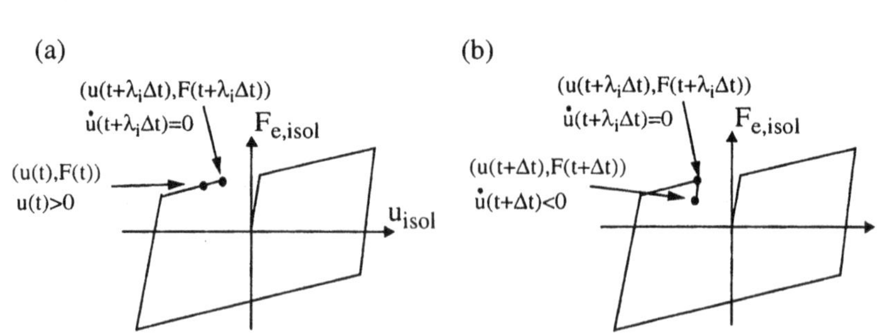

Figure B.4: Linear interpolation procedure to determine zero velocity instant (cont).

The time history analysis then continues with updated stiffness and damping for the reversed path. The damping is updated using Equations B.2 and B.3, once the stiffness has been updated as described above.

Finally, it may also be possible that both the yield force is exceeded and the velocity sign is reversed during a newly computed step $t + \Delta t$. In such a case, the earlier of the two events is identified by the algorithm, using the above linear interpolation procedures to determine which event occurs first.

During the time history analysis steps the displacement is computed as

$$u(t + \Delta t) = \qquad \text{(B.5)}$$

$$\frac{M\left\{\dfrac{4\ u(t)}{\Delta t^2} + \dfrac{4\ \dot{u}(t)}{\Delta t} + \ddot{u}(t) - \ddot{u}_g(t + \Delta t)\right\} + C(t)\left\{\dot{u}(t) + \dfrac{2\ u(t)}{\Delta t}\right\}}{\left(\dfrac{4\ M}{\Delta t^2} + \dfrac{2\ C(t)}{\Delta t} + K(t)\right)}$$

The time step during the interpolations equals $\lambda\Delta t$ instead of Δt. The velocity and acceleration may then be determined using Equations A.3 and A.5.

B.2 Multi Degree of Freedom Systems

Consider now the case of a building modeled as a shear beam with lumped masses at the floors and local non-linearities at the isolation level due to the bi-linear behavior of the isolation system. The model is shown in Figure B.5.

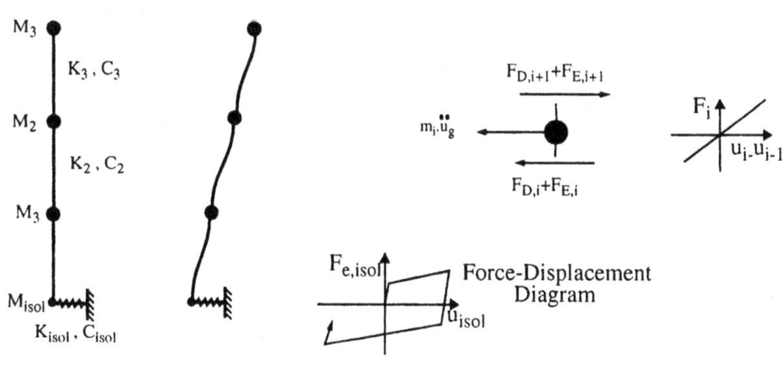

Figure B.5: Bilinear multi-DOF system.

The equations of motion of such a bilinear multi-DOF system may be written as

$$\left[F_I\right] + \left[F_D\right] + \left[F_E\right] = \left[0\right] \tag{B.6}$$

where at time $(t + \Delta t)$ we have the following forces:

$$\text{Inertia forces: } [F]_I(t + \Delta t) = [M]\ [\ddot{U}](t + \Delta t) \tag{B.7}$$

$$\text{Damping forces: } [F_D(t + \Delta t)] = [C(t + \Delta t)]\ \left[\dot{U}(t + \Delta t)\right] \tag{B.8}$$

$$\text{Deformation forces: } [F_E(t + \Delta t)] = [K(t + \Delta t)]\ (\left[U(t + \Delta t)\right] - [U(t)]) \tag{B.9}$$

For the 4-DOF model shown above, the system matrices are

$$\left[M\right] = \begin{bmatrix} m_{isol} & 0 & 0 & 0 \\ 0 & m_1 & 0 & 0 \\ 0 & 0 & m_2 & 0 \\ 0 & 0 & 0 & m_3 \end{bmatrix} \tag{B.10}$$

$$\left[C(t + \Delta t)\right] = \begin{bmatrix} (C_{isol}(t + \Delta t) + c_1) & -c_1 & 0 & 0 \\ -c_1 & (c_1 + c_2) & -c_2 & 0 \\ 0 & -c_2 & (c_2 + c_3) & -c_3 \\ 0 & 0 & -c_3 & c_4 \end{bmatrix} \tag{B.11}$$

$$\left[K(t + \Delta t)\right] = \begin{bmatrix} (k_{isol}(t + \Delta t) + k_1) & -k_1 & 0 & 0 \\ -k_1 & (k_1 + k_2) & -k_2 & 0 \\ 0 & -k_2 & (k_2 + k_3) & -k_3 \\ 0 & 0 & -k_3 & k_4 \end{bmatrix} \tag{B.12}$$

The equations of motion may therefore be written in the form

$$\begin{aligned} &\left[M\right]\ \left[\ddot{U}_{tot}(t)\right] + \left[C(t + \Delta t)\right]\ \left[\dot{U}(t)\right] + \\ &F_E(t) + \left[K(t + \Delta t)\right]\ (\left[U(t + \Delta t)\right] - \left[U(t)\right]) = \left[0\right] \end{aligned} \tag{B.13}$$

Substituting the expressions for the velocity and the acceleration assumed in the Newmark method previously described, we obtain

$$[M]\left(\frac{4}{\Delta t^2}\left(\left[U(t+\Delta t)\right]-\left[U(t)\right]-\left[\dot{U}t\right]\,\Delta t-0.25\,\Delta t^2\,\left[\ddot{U}(t)\right]\right)\right)+ \tag{B.14}$$

$$[C(t+\Delta t)]\left(2\,\frac{\left(\left[U(t+\Delta t)\right]-\left[U(t)\right]\right)}{\Delta t}-\left[\dot{U}t\right]\right)+F_E(t)+$$

$$[K(t+\Delta t)]\,\left(\left[U(t+\Delta t)\right]-\left[U(t)\right]\right)=-\left[M\right]\,\left[e\right]\,\ddot{U}_g(t+\Delta t)$$

This may be used to solve for the displacement at time $(t+\Delta t)$:

$$\left[U(t+\Delta t)\right]=\left\{\frac{4}{\Delta t^2}\left[M\right]+\frac{2}{\Delta t}\left[C(t+\Delta t)\right]+\left[K(t+\Delta t)\right]\right\}^{-1} \tag{B.15}$$

$$\left\{\left[K(t+\Delta t)\right]\,\left[U(t)\right]-F_E(t)+C\left(\left[\dot{U}t\right]+\frac{2}{\Delta t}\left[Ut\right]\right)\right.$$

$$\left.-\left[M\right]\,\left[e\right]\,\ddot{U}_g(t+\Delta t)+\left[M\right]\left(\frac{4}{\Delta t^2}\left[U(t)\right]+\frac{4}{\Delta t}\left[\dot{U}(t)\right]+\left[\ddot{U}(t)\right]\right)\right\}$$

Equation B.15 may be used to obtain the solution to the problem by computing the displacement, velocity and acceleration at each step.

The issues of concern related to the bilinear behavior, i.e. the loading and unloading paths and the possibility of exceeding the yield force level, are addressed in the same way as has been explained for a single DOF system. The only difference is that the time integration involves matrices instead of single values.

The loading and unloading paths are specified by the sign of the velocity of the mass at the isolation level. A change in velocity sign indicates reversion of the loading or unloading paths, while if the yield force is exceeded, a reduction of stiffness of the isolation system, from the initial elastic stiffness $K_{isol,el}$ to the post-yield stiffness $K_{isol,pl}$, is implied.

An extended algorithm of the type described in Paragraph B.1 has been developed and implemented in order to compute the exact point of either zero velocity or yielding, using the previously described linear interpolation procedure.

When a change in the sign of velocity occurs, a smaller integration time step is assumed and the iteration is repeated using a corresponding linearly interpolated value of ground acceleration. The process is repeated until the velocity is approximately equal to zero. The time history analysis then continues with updated stiffness and damping matrices for the reversed path.

Similarly, when the force at the isolation level exceeds the yield force, subsequent iterations are performed to find the exact time instant at which the force is approximately equal to the yield force. Once the time of yielding has been determined, the time history analysis continues with the updated post-yield stiffness.

Finally it may also be possible that both the yield force is exceeded and the velocity sign is reversed. In such cases, the earlier of the two events is detected and considered first, by the developed automatic procedure.

References

[1] AASHTO, *"Guide Specifications For Seismic Isolation Design"*, American Association of State Highway Officials, 1991.

[2] AASHTO, *"Guide Specifications For Seismic Isolation Design"*, American Association of State Highway Officials, 1998.

[3] Ahmadi H., Muhr A. and Ruiz-Roso F., "Prediction of post-yield behavior of isolated structures", *Seismic Design Practice into the Next Century*, Balkema, pp. 407-416, 1998.

[4] Almazan J., De La Lerra J. and Inaudi A., "Modelling aspects of structures isolated with the frictional pendulum system", *Earthquake Engineering and Structural Dynamics*, Vol. 27, pp. 845-867, 1998.

[5] Anderson T.L., "Seismic Isolation for the Los Angeles County Fire Command and Control Facility", *Proceedings of Seismic Engineering*, ASCE Structures Congress.

[6] Applied Technology Council, *Tentative Provisions for the Development of Seismic Regulations for Building, Including Proposed Amendments*, Report No. ATC3-06, Applied Technology Council, Palo Alto, California, 1984.

[7] Architectural Institute of Japan, *Recommendation for the Design of Base Isolated Buildings*, Sub Committee of Base Isolated Structures, Tokyo, Japan.

[8] Architectural Institute of Japan, *Final Report of International Workshop on Recent Developments in Base-Isolation Techniques for Buildings*, Tokyo, Japan, 1992.

[9] Asano K., Matsuda S. and Umino R., "Dynamic Behavior of Base Isolation System Considering Vertical Component of Earthquake Excitation", *Proceedings of the Second World Conference on Structural Control*, pp. 1111-1118, 1998.

[10] Asher J., Volkinburg V., Mayes. R., Kelly T., Sveinsson B. and Hussain S., "Seismic Isolation of the USC University Hospital", *Proceedings of the Fourth U.S. National Conference on Earthquake Engineering*, Palm Springs, CA, May 1990.

[11] Augenti N. and Serino G., "Proposal of a Design Methodology for Base Isolated Structures", *International Meeting on Earthquake Protection of Buildings*, Ancona, Italy, pp. 203-214, 1991.

[12] Bailey J.S. and Allen E.W., "Historic Seismic Protection", *Civil Engineering*, September Volume, 1988.

[13] Barbat and Bozzo, "Seismic analysis of base isolated buildings", *Archives of Computational Methods in Engineering*, Vol.4, pp.153-192, 1997.

[14] Bathe Klaus-Jurgen, *Finite Element Procedures,* Prentice-Hall Inc., Eaglewood Cliffs, New Jersey, 1996.

[15] Baumann T., "Sensitivity of seismic response with regard to the computational modelling of the nonlinear characteristics of isolators", *Seismic, Shock and Vibration Isolation*, Chung H. and Ishida K. Editors, pp. 9-17, 1998.

[16] Bozorgnia Y., Mahin S. and Brady G., "Vertical response of twelve structures recorded during the Northridge earthquake", *Earthquake Spectra,* Vol. 14, No. 3, pp. 411-432, 1999.

[17] Buckle I.G. and Kelly J.M., "Properties of slender elastomeric isolation bearings during shake table studies of large-scale model bridge deck", *Joint Sealing and Bearing Systems for Concrete Structures,* American Concrete Institute, Vol. 1, pp. 247-269, 1986.

[18] Buckle I.G. and Mayes R.L., "Seismic Isolation: History, Application and Performance A World View", *Earthquake Spectra*, Vol.6, No.2, pp. 161-201, 1990.

[19] Burns and Roe Inc., *Seismic Isolation: Studies for Liquid Metal Fast Breeder Plants,* 1983.

[20] Ceccoli C. Mazzotti C. and Savola M., "Non-linear seismic analysis of base-isolated RC frame structures", *Earthquake Engineering and Structural Dynamics*, Vol. 28, pp. 633-653, 1999.

[21] Chopra A.K., *Dynamics of Structures*, Prentice Hall, pp. 683-702, 1995.

[22] Chung W., Yun C., Kim N. and Seo J., "Shaking table and pseudodynamic tests for the evaluation of the seismic performance of base isolated structures", *Engineering Structures*, Vol. 21, pp. 365-379, 1999.

[23] Clough R. and Penzien J., *Dynamics of Structures*, McGraw Hill Inc., Second Edition, 1993.

[24] Colaiuda A., Turroni P., Cornaggia L. and Castellani A., "ITER seismic isolation: seismic studies of the ITER Tokamak structures", *Seismic, Shock and Vibration Isolation*, Chung H and Ishida K. Editors, pp. 209-213, 1998.

[25] Constantinou M., Caccese J. and Harris H., "Teflon Bearings in Base Isolation II: Modeling", *Journal of Structural Engineering*, Vol. 116, No 2, February, 1990.

[26] Constantinou M.C. and Reinhorn A.M., "Seismic Isolation and Control", *Computer Analysis and Design of Earthquake Resistant Structures: A Handbook*, Beskos D.E. and Anagnostopoulos S.A. Editors, CMP, Southampton, pp.839-867, 1997.

[27] Connor J.J. Klink B.S.A., *Introduction to Motion Based Design*, CMP, Southampton, 1996.

[28] DeLuca A., Faella G. and Mele E., "Serviceability and ultimate performance of base-isolated steel frames", *Behavior of Steel Structures in Seismic Areas*, E. & F.N. Spon., pp. 659-668, 1994.

[29] Dynamic Isolation Systems (DIS), *Performance of the Eel River Bridge in the Petrolia Earthquakes of April 25-26, 1992*, 1992.

[30] Dynamic Isolation Systems (DIS), *Seismic Isolation Update*, 1993.

[31] Dynamic Isolation Systems (DIS), *Seismic Isolation Update*, 1994.

[32] Dynamic Isolation Systems (DIS), *Seismically Isolated Buildings in North America*, revised 1994.

[33] Dynamic Isolation Systems (DIS), *Earthquake Performance of Structures on Lead-Rubber Isolation Systems*, Summary report by Dynamic Isolation Systems, revised September 1996.

[34] Eisenberg J. and Smirnov V., "Inexpensive seismoisolation and response control in Russia: recent developments", *Seismic, Shock and Vibration Isolation*, Chung H. and Ishida K. Editors, pp. 43-48, 1998.

[35] Elsesser E., Honeck W. and Walters M., "Seismic Retrofit of the Oakland City Hall, Oakland", *Structural Engineering International*, pp. 11-13, 1994.

[36] FEMA, *NEHRP Guidelines for Seismic Rehabilitation of Buildings*, National Earthquake Hazards Reduction (NEHRP), FEMA-273, 1998.

[37] FEMA, *Commentary on the NEHRP Guidelines for Seismic Rehabilitation of Buildings*, National Earthquake Hazards Reduction Program (NEHRP), FEMA-274, 1998.

[38] Forni M., Martelli A., Simoncini M., La Grotteria M., Dusi A., Bettinali F., Bonacina G. and Bergamo G., *"Progress od activities performed in Italy by ENEA and ENEL on seismic isolation and passive energy dissipation"*, *Seismic, Shock and Vibration Isolation*, Chung H. and Ishida K. Editors, pp. 19--41, 1998.

[39] Fukda S., Muramatsu Y., Katoh N., Masahiro N., Nisikawa I., Yamahuchi S. and Kani N., "Experimental study of the effects of center hole size and shear modulus of rubber on performance of natural rubber bearings", *Seismic, Shock and Vibration Isolation*, Chung H. and Ishida K. Editors, pp. 103-110, 1998.

[40] Fujita T., "Seismic Isolation and Response Control for Nuclear and Non-Nuclear Structures", *11th International Conference on structural Mechanics in Reactor Technology (SMiRT 11)*, Tokyo, Japan, 1991.

[41] Gifford M., "Strengthening parliament", *NZEngineering*, June Volume, pp. 29-30, 1993.

[42] HITEC, *Guidelines for the Testing of Seismic Isolation and Energy Dissipating Devices*, prepared by the Highway Innovative Technology Evaluation Center (HITEC), Civil Engineering Research Foundation (CERF) report, 1996.

[43] HITEC, *Evaluation findings for Enidine, Inc. Viscous Damper*, prepared by the Highway Innovative Technology Evaluation Center (HITEC), Civil Engineering Research Foundation (CERF) report, 1998.

[44] HITEC, *Evaluation findings for Oiles Multiple Viscous Shear Damper (MS-Damper)*, prepared by the Highway Innovative Technology Evaluation Center (HITEC), Civil Engineering Research Foundation (CERF) report, 1998.

[45] HITEC, *Evaluation findings for Taylor Devices Fluid Viscous Damper*, prepared by the Highway Innovative Technology Evaluation Center (HITEC), Civil Engineering Research Foundation (CERF) report, 1998.

[46] HITEC, *Evaluation findings for Dynamic Isolation Systems Elastomeric Bearings*, prepared by the Highway Innovative Technology Evaluation Center (HITEC), Civil Engineering Research Foundation (CERF) report, 1998.

[47] HITEC, *Evaluation findings for Earthquake Protection Systems, Inc. Friction Pendulum Bearings*, prepared by the Highway Innovative Technology Evaluation Center (HITEC), Civil Engineering Research Foundation (CERF) report, 1998.

[48] HITEC, *Evaluation findings for FIP-Energy Absorption Systems, L.C.C. Slider Bearings*, prepared by the Highway Innovative Technology Evaluation Center (HITEC), Civil Engineering Research Foundation (CERF) report, 1998.

[49] HITEC, *Evaluation findings for R.J. Watson, Inc. Sliding Isolation Bearings*, prepared by the Highway Innovative Technology Evaluation Center (HITEC), Civil Engineering Research Foundation (CERF) report, 1998.

[50] HITEC, *Evaluation findings for Scougal Rubber Corporation High Damping Rubber Bearings*, prepared by the Highway Innovative Technology Evaluation Center (HITEC), Civil Engineering Research Foundation (CERF) report, 1998.

[51] HITEC, *Evaluation findings for Skellerup Base Isolation Elastomeric Bearings*, prepared by the Highway Innovative Technology Evaluation Center (HITEC), Civil Engineering Research Foundation (CERF) report, 1998.

[52] HITEC, *Evaluation findings for Tekton Inc. Roller Bearings*, prepared by the Highway Innovative Technology Evaluation Center (HITEC), Civil Engineering Research Foundation (CERF) report, 1998.

[53] HITEC, *Evaluation findings for Tekton Inc. Steel Rubber Bearings*, prepared by the Highway Innovative Technology Evaluation Center (HITEC), Civil Engineering Research Foundation (CERF) report, 1998.

[54] Kani N. and Manabu O., "On AIJ Recommendation for the Design of Base Isolated Buildings", *International Workshop on Recent Developments in Base Isolated Techniques for Buildings*, Tokyo, pp. 3-14, 1992.

[55] Kawashima K. and Shoji Gaku, "Nonlinear Hysteretic Interaction between Device and Pier in an Isolated Bridge", *Proceedings of the Second World Conference on Structural Control*, 1998, pp. 876-884.

[56] Kelly J.M., "Seismic Base Isolation: Review and Bibliography", *Soil Dynamic and Earthquake Engineering*, Vol.6, pp. 202-206, 1986.

[57] Kelly J.M., "Base Isolation: Linear Theory and Design", *Earthquake Spectra*, Vol.6, No.2, pp. 223-243, 1990.

[58] Kelly J.M., *Earthquake Resistant Design with Rubber*, Springer-Verlag, Second Edition, 1997.

[59] Kelly J.M., "Recent Developments on Isolation of Civil Buildings in the United States", *International post-SMiRT Conference Seminar*, Capri, Italy, 1993.

[60] Kelly J. M., "The role of damping in seismic isolation", *Earthquake Engineering and Structural Dynamics*, Vol. 28, pp. 3-20, 1999.

[61] Kelly J. M., *Base isolation: origins and development*, National Information Service for Earthquake Engineering, On-line provided information, *http://nisee.ce.berkeley.edu/lessons/kelly.html*, November 1999.

[62] Kernigham B. and Ritchie D., *The C Programming Language*, Second Edition Prentice-Hall Software Section, Eaglewood Cliffs, New Jersey, 1988.

[63] Kitagawa Y., *"Base-Isolated Building Structures in Japan"*, Seminar on Seismic Design, Chile, 1989.

[64] Komodromos P. *Application of Seismic Isolation as a Performance-Based Earthquake-Resistant Design Method*, SM Thesis, Department of Civil and Environmental Engineering, MIT, 1995.

[65] Kulak R., Coveney V. and Jamil S., "Recovery characteristics of high damping elastomers used in seismic isolation bearings", *Seismic, Shock and Vibration Isolation*, Chung H. and Ishida K. Editors, pp. 79-86, 1998.

[66] Laskari B. and Kircher C., *"Evaluation of SEAOC/UBC Analysis Procedures"*, *Applied Technology Council 17-1*, Technical Papers on Seismic Isolation, pp. 149-160.

[67] Lew M. and Hudson M., "The effects of vertical ground motion on base-isolated building systems", *Earthquake Spectra*, Vol. 15, No. 2, pp 371-375, 1999.

[68] Lin A. N. and Shenton H.W., "Seismic performance of fixed-base and base-isolated steel frames", *Journal of Engineering Mechanics*, American Society of Civil Engineers (ASCE), Vol. 118, No. 5, pp.921-941, 1992.

[69] Makris N., "Near-Source earthquake, base isolated structures and semi-active dampers", *Earthquake Resistant Engineering Structures*, CMP, pp. 291-300, 1996.

[70] Malangone P. and Febraioli M., "A modal procedure for seismic analysis of non-linear base-isolated multistorey structures", *Earthquake Engineering and Structural Dynamics*, Vol. 27, pp. 397-412, 1998.

[71] Mayes R.L., Jones L.R. and Kelly T.E., "Impediments to the Implementation of Seismic Isolation", *Earthquake Spectra*, Vol.6, No.2, pp. 283-296, 1990.

[72] Mayes R.L., *"Impediments to the Implementation of Seismic Isolation"*, updated previous paper, Dynamic Isolation Systems, 1993.

[73] Mayes R.L., "Current Status and Future Needs of U.S. Seismic Isolation Codes", *International Workshop on Recent Developments in Base-Isolation Techniques for Buildings*, Tokyo, pp. 321-334, 1992.

[74] Mayes R.L., Jones L.R., Kelly T.E. and Knight R., "An Alternative Approach to the Distribution of Lateral Loads in Bridges", *Proceedings ACI Third World Congress on Joints and Bearings*, 1991.

[75] Mayes R.L., Jones L.R. and Kelly T.E., "The Economics of Seismic Isolation in Buildings", *International Workshop on Recent Developments in Base-Isolation Techniques for Buildings*, Tokyo, pp. 33-43, 1992.

[76] Mayes R.L., Jones L.R., Kelly T.E. and Button M.R., "Design Guidelines for Base-Isolated Buildings with Energy Dissipators", *Earthquake Spectra*, Vol.1, No.1, pp. 41-73, 1984.

[77] McKay G.R., Chapman H.E. and Kirkcaldie D.K.,"Seismic Isolation: New Zealand Applications", *Earthquake Spectra*, Vol.6, No.2, pp. 203-222, 1990.

[78] Mele E. and DeLuca A., "State of the art report on base isolation and energy dissipation", *Behavior of Steel Structures in Seismic Areas*, E. & F.N. Spon., pp. 631-658, 1994.

[79] Mizukoshi K., Yasaka A., Iizuka M. and Takabayashi K. "Failure Test of Laminated Rubber Bearings With Various Shapes", *Earthquake Engineering*, Tenth World Conference, pp. 2277-2286, 1992.

[80] Mokha A., Constantinou M. and Reinhorn A., "Teflon Bearings in Base Isolation I: Testing", *Journal of Structural Engineering*, Vol. 116, No 2, February, 1990.

[81] Mokha A., Navinchandra A., Constantinou M. and Zayas V., "Seismic isolation retrofit of large historic building", *Journal of Structural Engineering*, pp. 298-308, March 1996

[82] Moehle J.P., "Preliminary Report on the Seismological and Earthquake Engineering Aspects of the January 17, 1994 Northridge Earthquake", Report No. UCB/EERC-94/01, Earthquake Engineering Research Center, College of engineering, University of California at Berkeley, 1994.

[83] Moehle J.P., *Material on Hyogo-ken Nanbu earthquake*, Information released by Earthquake Engineering Research Center, College of Engineering, University of California at Berkeley, 1995.

[84] Monti G. Pint P. and Nuti C., "Response of conventional and isolated bridges under non-synchronous seismic motion", *European Seismic Design Practice*, Balkema, pp. 117-124, 1995.

[85] Mori A., Moss P., Cooke N. and Carr A., "The behavior of bearings used for seismic isolation under shear and axial load", *Earthquake Spectra*, Vol. 15, No. 2, pp 199-224, 1999.

[86] Mori A., Moss P., Cooke N. and Carr A., "The behavior of bearings used for seismic isolation under rotation and axial load", *Earthquake Spectra*, Vol. 15, No. 2, pp 225-244, 1999.

[87] Nabih Youssef, "Tall Order", *Civil Engineering*, pp.58-61, October 1999.

[88] Naeim F. and Kelly J.M., *Design of Seismic Isolated Structures*, John Wiley & Sons Inc., 1999.

[89] Nagarajaiah S., Reinhorn A.M., and Constantinou M.C., "Nonlinear dynamic analysis of 3D base-isolated structures", *Journal of Structural Engineering*, ASCE, Vol. 117, No. 7, pp. 2035-2068, September 1991.

[90] Nagarajaiah S., "Stability of elastomeric isolation systems", *Journal of Structural Engineering*, ASCE, Vol. 125, No. 9, pp. 946-954, September 1999.

[91] Nakagawa S. and Kawamura M., "Aseismic Design of C-1 Building: The Biggest Base Isolated Building in the World", *Post SMiRT11 Seminar, Seismic Isolation of Nuclear and Non-Nuclear Structures*, Japan, 1991.

[92] Olariou I., Olariou F. and Sarbu D., "Base isolation for seismic retrofitting of existing structures", *Seismic Design Practice into the Next Century*, Balkema, pp. 417-424, 1998.

[93] OSHPD, *An Acceptable Procedure For the Design and Renew of California Hospital Buildings Using Base Isolation*, Office of Statewide Health Planning and Development, State of California, 1989.

[94] OSHPD, *Division III: Earthquake Regulations for Seismic-Isolated Structures*, Office of Statewide Health Planning and Development, State of California, 1996.

[95] Park J. and Hisanori O., "Optimal yield level of bilinear seismic isolation devices", *Earthquake Engineering and Structural Dynamics*, Vol. 28, pp. 941-955, 1999.

[96] Paz Mario, *Structural Dynamics: Theory and Computation*, Third Edition, Van Nostrand Reinhold, New York, 1991.

[97] Robinson W.H., "Lead-Rubber Hysteretic Bearings Suitable for Protecting Structures During Earthquakes", *Earthquake Engineering and Structural Dynamics*, Vol.10, pp. 593-604, 1982.

[98] Roeder C.W., Stanton J.F. and Taylor A.W., *Performance of Elastomeric Bearings*, National Cooperative Highway Research Program (NCHRP) Report No.298, Transportation Research Board, Washington D.C, 1987.

[99] Stanton J. and Roeder C., "Advantages and Limitations of Seismic Isolation", *Earthquake Spectra*, Vol.7, No.2, pp. 301-323, 1991.

[100] SEAOC, "General Requirements for the Design and Construction of Seismic Isolated Structures", Structural Engineers Association of California, *Blue Book*, Appendix to Chapter 1, 1989.

[101] Shenton H.W. and Lin A.N., "Relative Performance of Fixed-Base and Base-Isolated Concrete Frames", *Journal of Structural Engineer*, Vol.119, No.10, pp. 2952-2967, 1993.

[102] Shenton H. and Hampton F., "Seismic response of isolated elevated water tanks", *Journal of Structural Engineering*, pp. 965-976, September 1999

[103] Skinner R.I., Robinson W.H. and McVerry G.H., *An Introduction to Seismic Isolation*, DSIR Physical Sciences, Wellington, New Zealand, John Wiley & Sons, 1993.

[104] Stanton J.F. and Roeder C.W., *Elastomeric Bearings Design, Construction and Materials*, National Cooperative Highway Research Program (NCHRP) Report No.248, Transportation Research Board, Washington D.C., 1984.

[105] Stewart J. Conte J. and Aiken I., "Observed behavior of seismically isolated buildings", *Journal of Structural Engineering*,, pp. 955-964, September 1999.

[106] Takaoka E., Masao I., Takenaka Y. and Yoshikawa K., "Base isolation system consisting of sliding and laminated rubber bearings", *Seismic, Shock and Vibration Isolation,* Chung H. and Ishida K. Editors, pp. 49-55, 1998.

[107] Taylor A.W., Lin A.N. and Martin J.W., "Performance of elastomers in Isolation Bearings: A literature Review", *Earthquake Spectra,* Vol.8, No.2, pp. 279-297, 1992.

[108] Uniform Building Code (UBC), "Earthquake Regulations for Seismic-Isolated Structures", Appendix to Chapter 23, pp. 875-891, 1991.

[109] Uniform Building Code (UBC), "Earthquake Design", Uniform Building Code, Part III, pp. 156-195, 1991.

[110] Uniform Building Code (UBC), "Earthquake Regulations for Seismic-Isolated Structures", Appendix to Chapter 16, 1994.

[111] Uniform Building Code (UBC), "Earthquake Regulations for Seismic-Isolated Structures", Appendix Chapter 16, Division IV, 1997.

[112] Vakakis A., Kounadis A. and Raftoyiannis I., "Use of non-linear localization for isolating structures from earthquake-induced motions", *Earthquake Engineering and Structural Dynamics*, Vol. 28, pp. 21-36, 1999.

[113] Yabana S., Kazuta H., and Ishida K., "Outlines and distinctive features of design guidelines for seismically isolated nuclear power facilities in Japan", *Seismic, Shock and Vibration Isolation*, Chung H. and Ishida K. Editors, pp. 229-235, 1998.

[114] Yamazaki N., Nakauchi H. and Miyazaki M., "Durability of Rubber Isolators-Characterization of a 100 Years Used Rubber Bearing", *International Workshop on Recent Developments in Base Isolated Techniques for Buildings*, Tokyo, pp. 211-213, 1992.

Index

A
advantages of seismic isolation 20
applications of seismic isolation 20

B
backup safety systems 90
backup stoppers 91
ball-bearing system 71
base isolation 2
basic components of
seismic isolation 19
behavior of elastomeric bearings 57
behavior of lead rubber bearings............ 41
bridge seismic isolation 92,93,95,109
buckling load 55
bulk modulus 51,52
bumpers 91

C
California Transportations (Caltrans) 58
Civil Engineering Research Foundation
(CERF) 59
clearance around the structure,
seismic gap 88
code requirements 97
comparison of a fixed-supported and
a seismically isolated structure 22
configuration of seismic isolation
system 87
connectivity details for utilities crossing
the isolation level 89
Constraints in implementation 26
conventional earthquake resistance
design 11
conventional earthquake resistance
design; philosophical consequences....... 17
cover rubber 38
crystallization, of elastomer due to low
temperatures 47

D
design basis earthquake 98
design spectra 12
disadvantages of conventional design 13
dynamic analysis procedure, UBC 104

E
EPDM 47
earthquake excitations,
used in the book 14
economics of seismic isolation 30
Eel River Bridge 85
effective modulus of elasticity 54
elastomeric isolation systems 37
evaluation tests of seismic isolation and
energy dissipating devices 58
existing buildings seismic
isolation codes 109

F
factors which enable the use of
seismic isolation 29
feasibility study for the use of
seismic isolation 140
Federal Highway Administration
(FHWA) 58
fiber-reinforced elastomeric bearings..... 58
Foothill Communities Law and
Justice Center 63
friction pendulum systems (FPS) 36
friction, coefficient of 35
full-scale dynamic tests of seismic
isolation devices 59
fundamental period influence 14

G
general requirements for the design
and construction of seismic isolated
structures, SEAOC 97

general UBC requirements 108
Gruppo de Lavoro Isolamento Sismico (GLIS) 77

H

Hanshin Expressway 84
high cost of elastomer 38
high damping rubber bearings (HDRBs) 8,43
high damping elastomeric bearings, first world application 63
high-rise buildings, pull-out forces on isolators 58
high tensile strength, fiber reinforced elastomeric bearings 58
Highway Innovative Technology Center (ITEC) 58
horizontal stiffness dependence on axial load 57
hospitals, seismic isolation codes 108

I

initial rigidity, provided by LRBs 39
isolation gap 89
Italy 78

J

Japan, seismic isolation regulations 109

K

Kobe earthquake 79,81

L

lack of extensive use of seismic isolation 32
large elastomeric bearings 48
lead 40
lead high damping rubber bearings (LHDRBs) 44
lead rubber bearings (LRBs) 8,39,40
limitations of seismic isolation 27
Loma Prieta earthquake 66,68
Los Angeles City Hall 69
low temperature elastomer stiffening 47
LRBs, stiffness of 41

M

maintenance and management of the isolation system 92,96
maximum credible earthquake 98
maximum displacement, bearing's plan area 55
maximum relative displacements 15
Ministry of Construction of Japan 71
modal contributions variation with degree of isolation 114
motivation to use seismic isolation 1
mounting steel plates 38

N

natural rubber bearings (NRBs) 39
neoprene 47
New Zealand Parliament Buildings 76
nitrile butadiene rubber 47
Northridge earthquake 65,79

O

Oakland City Hall 66
objectives and design philosophy of conventional design 11
oil and chemical substances resistance ... 48
ozone cracking 48

P

peak ground acceleration 141
pipes and cable connections, crossing the isolation level 89
Poisson ratio for rubber 53
polyisoprene 47
preliminary design methodology 139
pull-out forces on the bearings 58

R

Rangitaiki River Bridge, Te Teko 85
recommendation for the design of base isolated buildings, Japanese subcommittee guidelines 110
recommended lateral design requirements and commentary, *Blue Book* 97
required seismic gap for seismic isolation 5
resonance problem for low- to medium-rise structures 13
response spectra 14
restoring force, for sliding bearings 36

S

Salt Lake City and County Building 64
San Francisco City Hall 68

seismic hazard and site characteristics related to seismic isolation 27
seismic isolation system types 35
seismic upgrading of existing bridges 62
seismic upgrading of existing structures 62
shape factors SI and S2 49
shear stiffness of an LRB 56
single DOF systems, response 16
SIP Complex, Ancona, Italy 78
sliding isolation systems 35
spectral pseudo-accelerations 15
stability check, buckling load 55
static analysis procedure, UBC 99
stoppers 91
strain hardening effects 44
superstructure characteristics related to seismic isolation 27
surrounding structures constraints 28

T
tensile stresses exerted on elastomeric bearings 58
tentative seismic isolation design requirements 97
tests of seismic isolation systems 58
typical response spectra 4

U
Uniform Building Code 98
Uniform Building Code objectives 1
United States, regulations 97
United States, seismic isolation applications 62
Upland earthquake 64
uplift of bearings 92

V
vertical stiffness of a bearing 53
volumetric change, of a rubber sheet 51

W
Wasatch fault 65
Wellington Central Police Station 75
West Japan Computer Center 81
West Japan Postal Computer Center 82
William Clayton Building 74

Y
yield stress of lead 40
Young's and shear moduli 48

www.ingramcontent.com/pod-product-compliance
Ingram Content Group UK Ltd.
Pitfield, Milton Keynes, MK11 3LW, UK
UKHW020813290926
15470UKWH00003B/1128

9 781853 128035